ALLE ZEIT WACH 1842

36. Colloquium der Gesellschaft für Biologische Chemie
18.–20. April 1985 in Mosbach/Baden

# Neurobiochemistry

## Selected Topics

Edited by
B. Hamprecht and V. Neuhoff

With 64 Figures

Springer-Verlag Berlin Heidelberg New York Tokyo

Professor Dr. Bernd Hamprecht
Physiologisch-Chemisches Institut
der Universität
Hoppe-Seiler-Straße 1
D-7400 Tübingen, FRG

Professor Dr. Volker Neuhoff
Max-Planck-Institut für
Experimentelle Medizin
Hermann-Rein-Straße 3
D-3400 Göttingen, FRG

ISBN-13: 978-3-642-70942-5 e-ISBN-13: 978-3-642-70940-1
DOI: 10.1007/978-3-642-70940-1

Library of Congress Cataloging-in-Publication Data. Gesellschaft für Biologische Chemie. Colloquium (36th : 1985 : Mosbach, Baden-Württemberg, Germany) Neurobiochemistry : selected topics. Includes index. 1. Neurochemistry–Congresses. 2. Nervous system–Congresses. 3. Neural receptors–Congresses. I. Hamprecht, Bernd, 1939– . II. Neuhoff, Volker. III. Title. [DNLM: 1. Biochemistry–congresses. 2. Neurochemistry–congresses. 3. Neurons–congresses. W3 GE382R 36 th / WL 102.5 G389 1985n] QP356.3.G47 1985 599'.01'88
85-31563


Softcover reprint of the hardcover 1st edition 1985



2131/3130-543210

# Preface

The field of the neurosciences is one of the most rapidly growing in present biological research. Its molecular aspects are dealt with by the discipline of neurobiochemistry. As the theme of the Mosbacher Colloquium, we chose this term rather than the term "neurochemistry", in order to stress the dynamic biochemical aspects of present molecular neurobiology and to avoid the flavor of being purely descriptive and "static", which is frequently associated with the term neurochemistry. This appears the more warranted, since the natural products and analytical chemistry phase of discovering the basic chemical components of the nervous system has passed its culmination. The period of assessment has laid the foundation for studying the dynamic interplay of the various chemical components in the actual biological operation of nervous tissue. Thus, neurobiochemistry is that part of the neurosciences which is dominated by the ways of thinking and the methodology of biochemistry.

For this Colloquium only topics were selected that deal with the biochemistry of neurons. Thus, we excluded from the agenda other neural cells such as glial cells (astrocytes, ependymal cells, oligodendrocytes), meningeal cells, and capillary endothelial cells. This restriction was applied for two reasons: (1) The time available for the meeting did not allow an extensive display of the whole spectrum of neurobiochemical research. (2) The biochemistry of neurons is far more advanced than that of any other cell type of the nervous system. It may, therefore, exert paradigmatic influence on the development of the biochemistry of the other cell types.

Within the premises of neuronal biochemistry it appeared reasonable to start with a session on the biochemistry of factors that regulate neuronal development and differentiation. All other reports then concern function of mature neurons. We have tried to follow the sequence of events during the operation of neurons, i.e., as presynaptic functions: synthesis and degradation of neurohormones, storage and release of neurotransmitters, ion channels involved in action potentials and as postsynaptic operational elements receptors for neurohormones. Two chapters focus on the molecular aspects of neural disease and are meant to demonstrate that understanding of the molecular basis of disease essentially results from (sometimes unexpected) findings obtained in basic research.

In two special chapters we draw attention to two important aspects of present neurobiochemistry, techniques, and goals. The chapter by Jim Jackson emphasizes the enormous potential that is available for neurobiochemists in recombinant DNA techniques. The closing chapter by Irwin Levitan carefully opens the view onto one of the most eminent goals of neurobiochemistry, the elucidation of the mechanisms of memory.

With this in mind, it becomes obvious that our society might receive vehement impacts from the neurosciences that would deeply affect individual and group behavior. We believe that no country can afford not to be knowledgeable in these fields, if only to be able to judge such impacts and to react properly to them. We therefore hope that, from among the approximately 220 participants, the Colloquium may have attracted to neurobiochemistry a number of gifted young scholars who were looking for a highly exciting and worthwhile field of science.

We should like to thank the authors for their excellent presentations at the Colloquium and for their valuable contributions to this volume. It is regrettable that two of the speakers did not provide us with a manuscript of their exciting lectures.

We should like to acknowledge with gratitude the financial support by the Deutsche Forschungsgemeinschaft, the Minister für Wissenschaft und Kunst des Landes Baden-Württemberg, the Bayer AG (Leverkusen), Boehringer Mannheim (Mannheim), C.H. Boehringer Sohn (Ingelheim), the Grünenthal GmbH (Aachen), the Hoechst AG (Frankfurt), the Knoll AG (BASF, Ludwigshafen), Dr. Madaus & Co. (Köln), E. Merck (Darmstadt) and the Schering AG (Berlin).

November 1985

B. HAMPRECHT
V. NEUHOFF

# Contents

# Contributors

You will find the addresses at the beginning of the respective contributions

# Gene Regulation by Nerve Growth Factor

L. A. Greene[1], J. M. Aletta[1], D. E. Burstein[2], S. A. Drexler[1], M. E. Greenberg[3], D. Leonard[3], and E. B. Ziff[3]

## Introduction and Background

NGF is a protein with many profound, well-described (Levi-Montalcini 1966) actions on developing sensory and sympathetic neurons. To study the mechanism of action of NGF, we have employed an NGF-responsive clonal cell line - designated PC12 - which was isolated from a transplantable rat adrenal pheochromocytoma (Greene and Tischler 1976). The properties and experimental advantages of the PC12 line have been reviewed (Greene and Tischler 1982). In the absence of NGF, PC12 cells resemble their presumed non-neoplastic counterparts, noradrenergic adrenal chromaffin cells. After exposure to NGF, the cells slowly (i.e., over a time course of days) take on the phenotypic properties of sympathetic-like neurons so that, for instance, they cease proliferation, extend long neurites and become electrically excitable (Greene and Tischler 1982). Since, unlike sympathetic neurons, PC12 cells respond to, but do not require NGF for survival, they have the experimental advantage that they can be compared before and after various times of exposure to NGF.

## Long-Term, Specific Regulation of Gene Transcription by NGF

Studies with the PC12 line have provided evidence that a number of its actions are dependent on transcription. One important example is the promotion of neurite outgrowth. Exposure of PC12 cells to low levels of a variety of inhibitors of RNA synthesis has been found to block NGF-dependent initiation of neurite outgrowth (Burstein and Greene 1978). This and other findings led to formulation of the "priming" model for the mechanism of NGF action (Burstein and Greene 1978) in which it was proposed that NGF promotes neurite outgrowth in part by selectively stimulating the transcription-dependent synthesis and accumulation of material required for the formation and maintenance of neuritic processes. Recent studies have focused on identifying the gene products which underlie priming. Comparison of the major proteins synthesized by PC12 cells before and after long-term NGF exposure have revealed few qualitative or quantitative changes (McGuire et al. 1978). This has suggested that the transcription-dependent effects that do occur are likely to involve relatively low-abundance regulatory molecules rather than more abundant structural proteins (McGuire et al. 1978). One set of cellular regulatory proteins that has been considered for such a functional role in priming are microtubule-

[1]Department of Pharmacology

[2]Department of Pathology

[3]Department of Biochemistry
New York University School of Medicine, 550 First Avenue, New York, NY 10016, USA

36. Colloquium - Mosbach 1985
Neurobiochemistry

associated proteins or MAPs. Evidence has now accrued that PC12 cells contain at least four different classes of MAPs whose levels are regulated by NGF. One group is known as the high-molecular weight (HMW) MAPs. NGF brings about a large transcription-sensitive increase in levels of a particular HMW MAP, MAP 1.2 (Greene et al. 1983). A group of MAPs known collectively as the tau MAPs appear to be a second class whose levels are significantly increase after long-term exposure of PC12 cells to NGF (Feinstein et al. 1984). The use of a cDNA probe has indicated that NGF also brings about a concomitant increase tau MAP mRNA in PC12 cells (Feinstein et al. 1984). A third group of NGF-regulated MAPs fall in a mol wt range of 64-80 KD. While these MAPs are phosphorylated in unprimed PC12 cells, long term NGF treatment appears to dramatically enhance their degree of phosphorylation (Burstein et al. 1985). The fourth type of MAP that increases in relative abundance after treatment with NGF has, as estimated by SDS-PAGE, an apparent $M_r$ of about 34,000 and has been detected by repeated cycles of co-polymerization and depolymerization with brain tubulin (Green and Greene 1983).

Several considerations lead to a probable functional role of regulated MAPs in NGF-promoted priming and neurite outgrowth. (1) Microtubules appear to play a critical role in the growth and maintenance of neurites (McKeithan and Rosenbaum 1984). (2) In vitro studies have revealed that MAPs can, in turn, enhance the formation and stability of microtubules (McKeithan and Rosenbaum 1984). (3) Microtubules in long-term NGF-treated PC12 cells are considerably more resistant to colchicine than are the microtubules in non-NGF-treated PC12 cells (Black and Greene 1982) and therefore appear to be significantly more stable. (4) NGF appears to shift the proportion of tubulin in PC12 cells which is polymerized into microtubules by about three-fold (Black and Kurdyla 1984). (5) At least several of the NGF-regulated MAP's are apparently preferentially localized in PC12 cell neurites versus cell bodies (J. Aletta, unpublished results). (6) Two agents, lithium ion (Burnstein et al. 1985) and forskolin, have now been found which selectively block both phosphorylation of the 64-80 kD MAP's *and* neurite outgrowth. These points support the proposal that NGF stimulates neurite outgrowth at least in part by promoting microtubule assembly and stability via selective transcription-dependent regulation of MAP synthesis and phosphorylation.

In addition to effects on MAPs, there is also evidence for transcription-dependent actions of NGF that are associated with the acquisition by PC12 cells of a neuron-like phenotype. Among these is regulation of acetylcholinesterase activity. This increases by severalfold in response to NGF over a time course of 3-4 days and is blocked by actinomycin-D (Greene and Rukenstein 1981). Another example is the NGF-inducible large external (NILE) glycoprotein, which, over a time course of several weeks, is increased in specific level by five fold in response to NGF (McGuire et al. 1978, Salton et al. 1983). The effect of NGF on this neuronal marker protein (Salton et al. 1983) is also suppressed by inhibitors of RNA synthesis (McGuire et al. 1978). The thy-1 glycoprotein is a third example of a protein whose level undergoes significant transcription-inhibitor-sensitive up-regulation by NGF in PC12 cultures (Richter-Landsberg et al. 1985). To further study the mechanism of thy-1 regulation by NGF, a specific cDNA probe (Hedrick et al. 1984) for this molecule has been recently employed. Preliminary evidence by means of Northern blot analyses indicates that the increase in thy-1 protein is paralleled by a substantial increase in its specific levels of mRNA (S.A. Drexler, unpublished results).

To uncover additional genes that might be regulated by NGF, we have recently begun to carry out direct comparison of the mRNA's synthesized by PC12 cells before and after long-term treatment with NGF. To achieve this, poly(A)-containing cytoplasmic RNA has been isolated from the cells (± NGF treatment) and used to construct large cDNA libraries (each containing 1-4 x $10^6$ clones). These libraries have been compared by cross-hybridization. Thus far, an initial screen of the libraries followed by Northern blot analysis has yielded several mRNA's that are enhanced in level and several that decrease in level after NGF treatment. This approach promises to be a powerful means to detect and characterize NGF-induced changes in gene transcription.

One important feature that has emerged regarding the long-term regulation of genes by NGF is its apparent specificity. In addition to NGF, PC12 cells can respond to a variety of signals including epidermal growth factor, insulin, phorbol esters, elevated $K^+$ and permeant cAMP analogues (see Greene and Tischler, 1982 for review). However, these agents do not appear to bring about long-term, differentiation-related responses that require transcription such as neurite outgrowth, and enhancement of levels of MAP1, NILE glycoprotein, acetylcholinesterase or thy-1.

## Rapid, Transient NGF-Induced Changes in Gene Transcription

Thus far, this essay has considered NGF-dependent changes in transcription that occur over a time course of tens of hours to days. Are there more rapidly occurring changes. Until recently, there was only one example of a short-latency, transcription-inhibitor sensitive action of NGF - the induction of ornithine decarboxylase (ODC) activity. Exposure of PC12 cells (Greene and McGuire 1978) or sympathetic neurons (MacDonnell et al. 1977) to NGF for 4-6 h brings about a 20-40-fold induction of ODC activity. This effect is transient in that beyond 6 h, the levels of ODC activity rapidly fall towards basal levels. In contrast to the long-term transcription-dependent actions of NGF, ODC induction is not specific to NGF; other agents such as EGF and dB-cAMP also bring about induction, although of lower magnitudes, over similar time courses (Huff et al. 1981). Very recently, further studies have sought to test whether NGF brings about additional rapid transcriptional events (Greenberg et al. 1985). This has been performed by exposing PC12 cells to NGF for short times (5-240 min), isolating their nuclei and using the latter in nuclear "runoff" transcriptional assays (Greenberg and Ziff 1984). In this assay, transcripts initiated in vivo are permitted to go to completion in vitro, but in the presence of radiolabeled UTP and cold NTP's. The labeled transcripts are then hybridized with specific cDNA probes that have been adsorbed onto nitrocellulose paper. The level at which a given gene is being transcribed is thus reflected by the amount of labelled RNA which hybridizes with its corresponding cDNA probe. Application of this approach has revealed that NGF dramatically and selectively enhances the transcription of several genes within minutes of its addition to PC12 cultures. The rates of transcription of the proto-oncogene c-*fos* (Verma, 1984) and of actin were increased within 5 min of NGF treatment and reached approximately 80-fold enhancement within 1/2 h. Transcription of the proto-oncogene c-*myc* (Abrams et al. 1982) was also greatly elevated, but over a somewhat more delayed time course; onset of the effect was detectable by 15 min but it did not reach a maximum until 1-2 h. Induction of ODC gene transcription was also observed, but in this case, the effect was not detectable until several hours after NGF treatment. Over the same time course, transcription of several other genes was found to be unaffected. One aspect of the rapid gene regulation was that these effects were transient. That is, it appears that the levels of transcription in each case return to near basal rates within several hours following their induction. Another important feature of the rapid gene regulation is that it is not specific to NGF. EGF, which produces several rapid responses in PC12 cells comparable to those obtained with NGF, but which does not promote neuronal differentiation or long-term differentiation-related transcriptional changes (Huff et al. 1981), was found to induce a similar set of fast transcriptional changes in PC12 cultures. In addition, a number of other agents including insulin, fibroblast growth factor, phorbol ester, dibutyryl-cAMP and depolarizing levels of $K^+$ also brought about rapid induction of c-*fos* transcription in PC12 cultures.

## Conclusions

The above findings regarding gene regulation by NGF raise a number of intriguing questions and allow several conclusions to be drawn concerning both the role of NGF in development and its mechanism of action

1. It is now unequivocal that NGF brings about regulation of gene expression at the transcriptional level.
2. NGF appears to promote two distinct classes of gene regulation. One class is not exclusive to NGF and includes rapid, transient enhancement of the transcription of specific genes; a second class of gene regulation appears to occur after a greater delay, to be long lasting in duration and to be more exclusive to NGF.
3. Gene regulation by NGF, both fast and slow, is not temporally synchronized. For instance, effects on c-*fos* and actin precede those on c-*myc*. With respect to the slower, NGF-specific class of changes, thy-1 regulation is detectable before that of acetylcholinesterase which in turn precedes that of MAP1 and NILE glycoprotein. These observations raise the question as to how continuous exposure to NGF triggers a stereotyped and temporally desynchronized set of alterations in gene transcription.
4. The slower, specific changes in gene regulation appear to be associated with neuronal phenotypic differentiation. In some cases, as for regulation of MAPs, such gene regulation by NGF may lead to qualitative changes in phenotype (i.e., neurite outgrowth). In other cases, as for increases in acetylcholinesterase and NILE glycoprotein, slow effects on gene transcription in response to NGF may quantitatively modulate the degree of expression of neuronal properties.
5. The fast transcriptional responses to NGF, being nonspecific, cannot be *sufficient* to account for NGF-induced differentiation.
6. It is possible that the fast changes, if not sufficient, are *required* elements in NGF-promoted differentiation. For instance, it could be that such early responses permit the cells to undergo delayed, NGF-specific gene regulation. There are now several other systems in which proto-oncogene expression precedes differentiation (cf. Gonda and Metcalf 1984, Müller and Wagner 1984).
7. An alternative role for the early transcriptional responses is that they are involved in mitogenesis. Rapid mitogenic growth-factor enhanced transcription of proto-oncogenes have been observed in a number of other systems (cf. Greenberg and Ziff 1984). Since the expression of these genes appears to be related to oncogenic transformation (Bishop 1983), it has been suggested that effects on oncogenes may be necessary steps in stimulation of proliferation by growth factors (Greenberg and Ziff 1984). This could also be the case for NGF. NGF has been found to be a potent mitogen for

several mutant PC12-cell-derived lines that do not undergo NGF-promoted neuronal differentiation (Burstein and Greene 1982). In addition, evidence has been presented (Boonstra et al. 1983) that NGF transiently stimulates DNA synthesis in PC12 cultures before they undergo neuronal differentiation. Although NGF is unlikely to exert mitogenic actions on mature neurons, it is possible that the factor could serve to regulate proliferation of responsive neuroblasts.

In summary, evidence has been presented here that NGF regulates two distinguishable classes of gene expression. It remains to uncover the mechanism(s) whereby NGF affects gene expression and to clearly define the functional role of each of the affected genes.

*Acknowledgments.* This work was supported in part by grants from the March of Dimes Birth Defects Foundation, American Cancer Society (MV-75) and by NIH grants NS16036, and GM30760. We thank Y. Calderon for aid in preparation of this manuscript.

## References

Abrams HD, Rohrschneider LR, Eisenman RN (1982) Nuclear location of the putative transforming protein of avian myelocytomatosis virus. Cell 29:427-429

Bishop JM (1983) Cellular oncogenes and retroviruses. Annu Rev Biochem 52:301-354

Black MM, Greene LA (1982) Changes in colchicine susceptibility of microtubules associated with neurite outgrowth: studies with nerve growth factor-responsive PC12 pheochromocytoma cells. J Cell Biol 95:379-386

Black MM, Kurdyla JT (1984) Nerve growth factor (NGF)-induced changes in microtubules of PC12 cells. Trans Am Soc Neurochem 15:163

Boonstra J, Moolenaar WH, Harrison PH, Moed P, Van der Saag PT, de Laat SW (1983) Ionic responses and growth stimulation induced by nerve growth factor and epidermal growth factor in rat pheochromocytoma (PC12) cells. J Cell Biol 97:92-98

Burstein DE, Greene LA (1978) Evidence for both RNA-synthesis-dependent and -independent pathways in stimulation of neurite outgrowth by nerve growth factor. Proc Natl Acad Sci USA 75:6059-6063

Burstein DE, Greene LA (1982) Nerve growth factor has both mitogenic and anti-mitogenic actions. Dev Biol 94:477-482

Burstein DE, Seeley PJ, Greene LA (1985) Lithium ion inhibits nerve growth factor (NGF)-induced neurite outgrowth and phosphorylation of NGF-modulated microtubule-associated proteins. J Cell Biol 101:862-870

Feinstein S, Drubin D, Sherman-Gold R, Kirschner M, Shooter EM (1984) Mobilization of cytoskeletal elements during NGF induced neurite outgrowth in PC12 cells. Soc Neurosci Abstr 10:163

Gonda TJ, Metcalf D (1984) Expression of *myb*, *myc* and *fos* proto-oncogenes during the differentiation of murine myeloid leukemia. Nature (London) 310:249-251

Green SH, Greene LA (1983) Increased low molecular weight microtubule-associated protein (MAP) in PC12 cells following long term exposure to NGF. Soc Neurosci Abstr. 9:207

Greenberg ME, Ziff EB (1984) Stimulation of 3T3 cells induces transcription of the c-*fos* proto-oncogene. Nature (London) 311:433-438

Greenberg ME, Greene LA, Ziff EB (1985) Nerve growth factor and epidermal growth factor induce rapid transient changes in c-*fos* and c-*myc* proto-oncogene expression in PC12 cells. J Biol Chem 260:14101-14110

Greene LA, McGuire JC (1978) Induction of ornithine decarboxylase by NGF dissociated from effects on survival and neurite outgrowth. Nature (London) 276:191-194

Greene LA, Rukenstein A (1981) Regulation of acetylcholinesterase activity by nerve growth factor: Role of transcription and dissociation from effects on proliferation and neurite outgrowth. J Biol Chem 256:6363-6367

Greene LA, Tischler AS (1976) Establishment of a noradrenergic clonal line of rat adrenal pheochromocytoma cells ehich respond to nerve growth factor. Proc Natl Acad Sci USA 73:2424-2428

Greene LA, Tischler AS (1982) PC12 pheochromocytoma cultures in neurobiological research. In: Federoff S, Hertz L (eds) Advances in cellular neurobiology, vol III. Academic Press, London New York, pp 373-414

Greene LA, Liem RKH, Shelanski ML (1983) Regulation of a high molecular weight microtubule-associated protein in PC12 cells by nerve growth factor. J Cell Biol 96:76-83

Hedrick SM, Cohen DI, Nielsen EA, Davis MM (1984) Isolation cDNA clones encoding T cell-specific membrane-associated proteins. Nature (London) 308:149-153

Huff K, End D, Guroff G (1981) Nerve growth factor induced alteration in the response of PC12 pheochromocytoma cells to epidermal growth factor. J Cell Biol 88:189-198

Levi-Montalcini R (1966) The nerve growth factor, its mode of action on sensory and sympathetic nerve cells. Harvey Lect 60:217-259

MacDonnell PC, Nagaiah K, Lakshmanan J, Guroff G (1977) Nerve growth factor increases activity of ornthinine decarboxylase in superior ganglia of young rats. Proc Natl Acad Sci USA 74:4681-4684

McGuire JC, Greene LA, Furano AV (1978) NGF stimulates incorporation of fucose of glucosamine into an external glycoprotein in cultured rat PC12 pheochromocytoma cells. Cell 15:357-365

McKeithan TW, Rosenbaum JL (1984) The biochemistry of microtubules: a review. In: Shay JW (ed) Cell and muscle motility, vol V. Plenum Press, New York, pp 255-288

Müller R, Wagner EF (1984) Differentiation of F9 teratocarcinoma stem cells after transfer of c-*fos* proto-oncogenes. Nature (London) 311:438-442

Richter-Landsberg C, Greene LA, Shelanski ML (1985) Cell surface thy-1-cross-reactive glycoprotein in cultured PC12 cells: Modulation by nerve growth factor and association with the cytoskeleton. J Neurosci 5:468-476

Salton SRH, Richter-Landsberg C, Greene LA, Shelanski ML (1983) The NGF-inducible large external (NILE) glycoprotein: Studies on a central and peripheral marker. J Neurosci 3:441-454

Verma IM (1984) From c-*fos* to v-*fos*. Nature (London) 308:317

# Implications of Proteases and Protease Inhibitors in Neurite Outgrowth

D. Monard [1]

## Introduction

There is increasing experimental evidence that the proteolytic activity associated with cells can strongly affect their behavior. The role of cell-surface protease activity in cell migration has been stressed in many developmental systems [1,2]. Migrating cells, including cells transformed by oncogenic viruses, have more cell-associated proteolytic activity than stationary, differentiated cells [3,4]. Cell-associated proteolytic activity is usually estimated by measuring cell-derived plasminogen activators [5]. Plasminogen activators are serine proteases that convert plasminogen to plasmin, which then degrades substrates such as fibrin [6] or casein [7]. In the nervous system, an increased amount of plasminogen activator can be attributed to cultured, cerebellar granule neurons if the dissociation and the initiation of the cultures take place at the developmental stage at which these cells are migrating [8]. It was also demonstrated that, in neuroblastoma cells, plasminogen activator activity can be preferentially localized at the growth cone when neurite outgrowth is induced by treatment with dibutyryl cyclic AMP [9]. These results have suggested that such proteolytic activity is required for the migratory behavior of the growth cone during neurite outgrowth [9].

## A Glia-Derived Neurite-Promoting Factor with Protease Inhibitory Activity

Cultured glial cells, including glioma cells, release a macromolecular factor which induces a dose-dependent neurite outgrowth in neuroblastoma cells [10]. This neurite-promoting activity is also detected in the medium conditioned by rat brain primary cultures if they are established at or after a critical developmental stage at which the burst of glial cell proliferation takes place [11]. These results indicate that glial cells release exogeneous signals which induce or at least contribute to the regulation of neurite outgrowth. This glia-derived neurite-promoting factor (GdNPF) is distinct from the well-studied nerve growth factor (NGF) [12].

During the purification of GdNPF, it became obvious that fractions with GdNPF activity also contained a very potent inhibitor of urokinase or plasminogen activator [13]. This led to the finding that GdNPF can be specifically adsorbed by urokinase immobilized on Sepharose beads. The small amount of protein which can be eluted from such a urokinase affinity chromatography possesses both neurite-promoting and protease inhibitory activities [14]. A procedure for isolating microgram amounts of

[1] Friedrich Miescher-Institut, P.O. Box 2543, CH-4002 Basel, Switzerland

36. Colloquium - Mosbach 1985
Neurobiochemistry

electrophoretically homogeneous GdNPF has recently been developed [15]. Using this pure material, it could be demonstrated that neurite-promoting and protease inhibitory activities are due to the same 43 kd protein. GdNPF inhibits both urokinase and plasminogen activator through the formation of a SDS-resistant, high-molecular-weight complex. GdNPF inhibits 0.1 mU urokinase at $6 \times 10^{-9}$ M and promotes neurite outgrowth at $6 \times 10^{-11}$ M [15]. Recent experiments have shown that pure GdNPF strongly inhibits the migration of granule cells monitored in cultured explants of the cerebellum, suggesting that surrounding glial cells have an active role regulating neuronal migration [16].

## Protease Inhibitors with Neurite-Promoting Activity

Many well-known serine protease inhibitors have been tested for their ability to promote neurite outgrowth [13]. Only hirudin and the tripeptide D-Phe-Pro-Arg$CH_2Cl$, known for their high specificity for thrombin-like proteases [17,18], are as potent as GdNPF, promoting neurite outgrowth at $5 \times 10^{-11}$ M. All the other inhibitors tested are either inactive or have only a marginal effect at high concentrations, with the exception of leupeptin, which can clearly promote neurite outgrowth at $10^{-4}$ M [13].

## Identity of the Proteases Associated with Neuronal Cells

GdNPF shows a dose-dependent inhibition of the two molecular weight forms of plasminogen activator associated with neuronal cells [19]. The plasminogen-dependent proteolytic activity released into the medium by mouse neuroblastoma cells is inhibited by 50% at $4 \times 10^{-10}$ M GdNPF. $9 \times 10^{-9}$ M GdNPF is necessary to cause a 50% inhibition of the same enzymatic activity when it is still associated with the intact neuroblastoma cells. $6.5 \times 10^{-10}$ M GdNPF will inhibit 50% of the same proteolytic activity when it is measured in neuroblastoma cell extract.

Is GdNPF acting solely by inhibiting plasminogen activator? At least 6 to 7 proteins can be labeled when neuronal membranes are incubated with $^3$H-diisopropylfluorophosphate, suggesting that several different serine protease are associated with neuronal cells [19]. $10^{-6}$ M D-Phe-Pro-Arg$CH_2Cl$ is required for 50% inhibition of the different forms of neuronal plasminogen activator mentioned above and hirudin does not inhibit this protease even at $10^{-4}$ M. These two inhibitors having thrombin specificity, their identical potency as GdNPF for the promotion of neurite outgrowth indicate that a protease with thrombin-like specificity is associated with the neuronal cells as well. Preliminary results indicate that GdNPF is also a very potent inhibitor for thrombin and trypsin [15], thus one has to consider that the inhibition of proteases distinct from plasminogen activator could as well lead to the biological effects of GdNPF. It will obviously be important to identify which of the proteases associated with the neuronal cells is preferentially inhibited by GdNPF. Moreover, it will be necessary to identify the function of each of these proteases, in either neuronal cell migration or neurite outgrowth. Finally, the substrate specificity for each of these proteases will also have to be determined to eventually elucidate the mode of action of GdNPF.

In view of this highly selective protease inhibitory activity, the most likely hypothesis is that GdNPF regulates neuronal cell migration and neurite outgrowth by controlling the activity of specific cellular proteases. This seems to contradict the hypothesis that proteolytic activity is required for the advancement of the growth cone in the surrounding extracellular matrix [9]. The following considerations indicate that this is only an apparent contradiction.

## Balance Between Proteases and Protease Inhibitors as a Regulatory Step of Neuronal Cell Migration

Proteolytic activity is one of the functions required for the migration of the neuronal cells. Specific proteases associated with migrating neuronal cells would be required to allow a cell to make its way through the extracellular matrix (ECM). The importance of an ECM component such as fibronectin for the migration of neural crest cells has already been demonstrated [20]. Cellular migration could thus require the degradation of some ECM components. The degradation of fibronectin during this cellular event has, in fact, been suggested but not yet clearly demonstrated [21]. On the other hand, if ECM components (fibronectin, laminin, collagen) and cellular adhesion molecules (NCAM, L1, NgCAM [22,23,24] function as guiding cues, then their proteolytic degradation would be a necessary contribution to the dynamics of the phenomenon by regulating the transitory nature of such interactions. Glial cells would then have an opportunity to influence neuronal cell migration by releasing inhibitors able to affect such degradative enzymes. The type and amount of inhibitor released could even change within the local environment of a migrating or differentiating cell, allowing the glial cells to influence the migration of each neuronal cell individually. In addition one can assume that the amount of proteolytic activity required for neuronal migration creates conditions which are incompatible with neurite outgrowth. The reduction in proteolytic activity caused by glia-derived protease inhibitors would therefore not only influence the extent of migration but, at the same time, create new conditions allowing the initial target-independent phase of neurite outgrowth.

## Balance of Proteases and Protease Inhibitors in the Regulation of Neurite Outgrowth

If the reduction of proteolytic activity leads to the cessation of cellular migration, then a certain amount of proteolytic activity has to remain localized at the growth cone to allow its progression [9]. Time-lapse cinematographic studies have illustrated the complexity and the speed of the filopodia motility at the growth cone. In other words, the growth cone can be considered as a migrating subcellular structure. It seems reasonable that the proteolytic activity is important for the movements of filopodia. It is likely that the organization of the cytoskeleton in the growing neurite establishes a preferential cytoplasmic transport system (an immature axoplasmic transport) which brings to the growth cone all the elements required for the dynamics of the filopodia. Proteolytic enzymes could as well follow such a pathway in order to become inserted and localized at higher concentrations at the level of the filopodia, where they would be required for the penetration of these structures through surrounding tissue. However, neurite extension would also require the stabilization of the structures which have found the appropriate cues and which are then to become the more rigid regions localized just behind the active growth cone. The glia-derived protease inhibitors would promote the stage in which a filopodia structure becomes stabilized as a neuritic structure (Fig. 1).

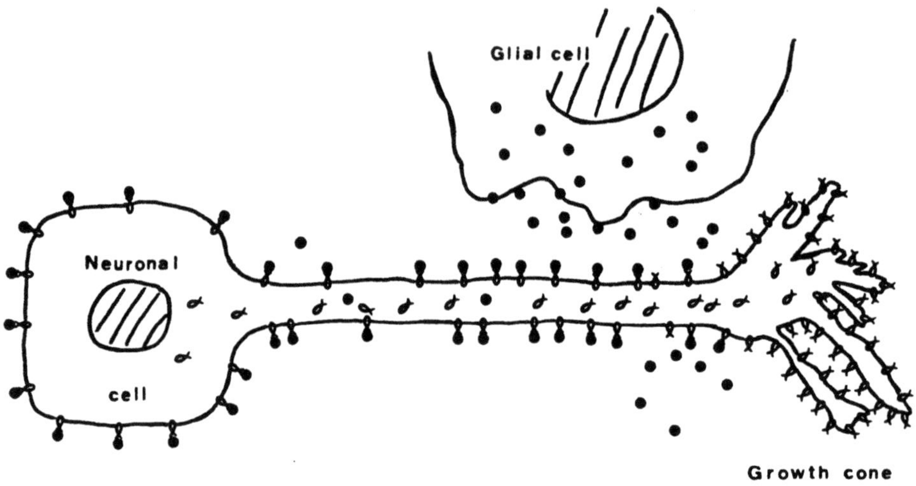


Fig. 1. Both proteolytic activity and inhibition of proteolytic activity are required for neurite outgrowth or regeneration. ∝: active proteases at the mobile filopodia of the growth cone. • : protease inhibitors provided by the surrounding glial cells or induced by a proper trigger (trophic factor) in the neuronal cell itself. ∝• : inactive protease-inhibitor complexes in the subcellular region requiring stabilization, e.g., the subgrowth cone structure

Such protease inhibitors would not have to be specifically provided by the surrounding glial cells. One can also envisage that, under certain conditions, certain neurons would, by themselves, produce an endogeneous amount of inhibitor which could also be transported, inserted, and then react with the filopida-associated proteases in order to contribute to the post-filopodia stabilisation. This possiblity is strongly supported by the recent report that NGF treatment, known to promote neurite outgrowth in PC12 pheochromocytoma cells, also induces the synthesis of a protein which appears to be identical to GdNPF. Like GdNPF, this protein is a 43 kd protease inhibitor which forms SDS-resistant complex with $^{125}$I urokinase [25,26]. These results suggest that the appearance of protease inhibitory activity might be an important step in the chain of events involved in the mode of action of NGF. If production of protease inhibitor can be stimulated by a specific trigger in the neuronal cell itself, one can envisage that the surrounding glial cells would have not an inductive but rather a permissive effect in contributing quantitatively to the amount of inhibitor required for a more efficient neurite extension. Alternatively, glia-derived protease inhibitors could preferentially control neuronal migration and onset of neurite outgrowth when neuron-derived protease inhibitors would mainly regulate neuritic growth in situ.

Finally, one can speculate that similar mechanisms are involved in regeneration phenomena. It is generally acknowledged that regeneration can take place in the peripheral nervous system (PNS) but not in the central nervous system (CNS). Recent transplantation experiments [27] have indicated that CNS neurons have the ability to regenerate their neurites if in the proper environment. The lesioned site in the PNS is thought to provide this appropriate environment. The recent report that a protein with the same characteristics as GdNPF is found after lesion in the PNS but not in the CNS [25,26] is an additional argument supporting the hypothesis that the inhibition of an excess of proteolytic activity is a key event in the regeneration of neurites.

## Conclusion

A fine and precisely localized balance between neuronal proteolytic activities and inhibitors, provided either by surrounding glial cells or by the properly stimulated neuroblast itself, is probably required for sustained neurite elongation. Such biochemical events could also be of crucial importance for regeneration phenomena. The purification and characterization of a glioma-derived 43 kd protease inhibitor offers the opportunity to produce the antibodies and molecular probes necessary to study the precise involvement of such molecules in neurite elongation and regeneration.

*Acknowledgments.* I thank Drs. A. Matus, F. Meins, and G. Thomas for critical reading of the manuscript.

## References


1. Sherman MI, Strickland S, Reich E (1976) Differentiation of early mouse embryonic and teratocarcinoma cells in vitro: plasminogen activator production. Cancer Res 36:4208-4216
2. Topp W, Hall JD, Marsden M, Teresky AK, Rifkin D, Levine AJ, Pollack R (1976) In vitro differentiation of teratomas and the distribution of creatine phosphokinase and plasminogen activator in teratocarcinoma-derived cells. Cancer Res 36:4217-4223
3. Valinsky JE, Reich E, Le Douarin NM (1981) Plasminogen activator in the bursa of Fabricius: correlations with morphogenetic remodeling and cell migration. Cell 25:471-476
4. Unkeless JC, Tobia A, Ossowski L, Quigley JP, Rifkin DB, Reich E (1973) An enzymatic function associated with transformation of fibroblasts by oncogenic viruses. J Exp Med 137:85-111
5. Ossowski L, Quigley JP, Reich E (1974) Fibrinolysis associated with oncogenic transformation. J Biol Chem 249:4312-4320
6. Williams JRB (1951) The fibrinolytic activity in urine. Br J Exp Pathol 32:530-537
7. Goldberg AR (1974) Increased protease levels in transformed cells: a casein overlay assay for the detection of plasminogen activator production. Cell 2:95-102
8. Krystosek A, Seeds NW (1981) Plasminogen activator secretion by granule neurons in cultures of developing cerebellum. Proc Natl Acad Sci USA 78:7810-7814
9. Krystosek A, Seeds NW (1981) Plasminogen activator release at the neuronal growth cone. Science 213:1532-1534
10. Monard D, Solomon F, Rentsch M, Gysin R (1973) Glia-induced morphological differentiation in neurobalstoma cells. Proc Natl Acad Sci USA 70:1894-1897
11. Schuerch-Rathgeb Y, Monard D (1978) Brain development influences the appearance of glia factor-like activity in rat brain primary cultures. Nature (London) 273: 308-309
12. Monard D, Stockel K, Goodman R, Thoenen H (1975) Distinction between nerve growth factor and glial factor. Nature (London) 258:444-445
13. Monard D, Niday E, Limat A, Solomon F (1983) Inhibition of protease activity can lead to neurite extension in neuroblastoma cells. Progr Brain Res 58:359-364
14. Monard D, Murato K, Guenther J (1985) Glia-induced neurite extension as a model to detect some biochemical interactions between glia and neurons. In: Dumont JE, Hamprecht B, Nunez J (eds) Hormones and cell regulation, vol IX. Elsevier Science Publ, Amsterdam, pp 391-398
15. Guenther J, Nick H, Monard D (1985) A glia-derived neurite-promoting factor with protease inhibitory activity. EMBO 4:1963-1966
16. Lindner J, Guenther J, Nick H, Zinser G, Antonicek H, Schachner M, Monard D (1985) Modulation of granule cell migration by a glia-derived protein. In preparation

17. Markwardt F, Walsmann P (1958) Die Reaktion zwischen Hirudin und Thrombin. Hoppe-Seyler's Z Physiol Chem 312:85-98
18. Kettner C, Shaw E (1979) D-Phe-Pro-Arg$CH_2Cl$ - A selective affinity label for thrombin. Thrombosis Res 14:969-973
19. Murato K, Monard D (1985) In preparation
20. Thiery JP, Duband JL, Delouve A (1982) Pathways and mechanisms of avian trunk neural crest migration and localization. Dev Biol 93:324-343
21. Thiery JP, Duband JL, Rutishauser U, Edelman GM (1982) Cell adhesion molecules in early chicken embryogenesis. Proc Natl Acad Sci USA 79:6737-6741
22. Edelman GM (1983) Cell adhesion molecules. Science 219:450-457
23. Lindner J, Rathjen FG, Schachner M (1983) L1 mono- and polyclonal antibodies modify cell migration in early postnatal mouse cerebellum. Nature (London) 305: 427-430
24. Grumet M, Edelman GM (1984) Heterotypic binding between neuronal membrane vesicles and glial cells is mediated by a specific cell adhesion molecule. J Cell Biol 98:1746-1756
25. Patterson PH (1985) On the role of proteases, their inhibitors and the extracellular matrix in stimulating neurite outgrowth. J Neurochem 44: Suppl s9B
26. Patterson PH (1985) J Physiol (Paris) in press
27. Benfey M, Aguayo AJ (1982) Extensive elongation of axons from rat brain into peripheral nerve grafts. Nature (London) 296:150-152

# Studies on the Cholinergic Differentiation Factor for Cultured Sympathetic Neurons

K. Fukada [1]

Over the past 10 years, it has been shown that individual neurons of neural crest origin are plastic with respect to transmitter phenotype (Patterson 1978, Le Douarin 1980). Transplantation studies in chick-quail chimeras have shown that the population of neural crest cells can become either noradrenergic or cholinergic, depending on the embryonic environment in which they are placed (Le Douarin 1980). The transmitter phenotype of the neurons can be switched even after the cells cease migrating and form ganglia (Le Douarin et al. 1978). Moreover, studies under the more defined conditions of cell culture have shown that the alteration of transmitter phenotype is not due to a selection of certain neural crest subpopulations for survival but a change in transmitter metabolism at the single-cell level (Reichardt and Patterson 1977, Potter et al. 1981).

The existence of a diffusible factor which controls the transmitter phenotype of these neurons without affecting neuronal survival or growth has been demonstrated in culture studies. When sympathetic neurons taken from neonatal rats are grown in the virtual absence of non-neuronal cells, they develop many of the properties expected of noradrenergic neurons in vivo with a time course similar to that observed in vivo (Claude 1973, Mains and Patterson 1973a,b,c, Rees and Bunge 1974, Landis 1980). On the other hand, when the same sympathetic neurons are grown in the presence of appropriate non-neuronal cells, in culture media conditions by them (CM), or on a monolayer of p-formaldehyde fixed heart cells (Hawrot 1980), they can be influenced to become cholinergic. These neurons develop the ability to synthesize acetylcholine (ACh), and form functional cholinergic synapses along a time course similar to that of noradrenergic neurons in culture (O'Lague et al. 1974, Patterson and Chun 1974, 1977a,b, Bunge et al. 1978, Patterson 1978) and they do so in a dose-dependent manner. The stronger the cholinergic stimulus given to the neurons, the more cholinergic the neurons become (Landis et al. 1976, Patterson and Chun 1977a), and at the same time many of the noradrenergic properties, including catecholamine (CA) synthesis and storage, and levels of enzymes involved in CA synthesis and degradation, are reduced (Johnson et al. 1976, Landis 1976, 1980, Johnson et al. 1980, Pintar et al. 1981, Swerts et al. 1983, Wolinsky and Patterson 1983).

The key question, then, is: Does this cholinergic factor demonstrated in culture studies play a physiological role in normal development in vivo? Recent studies by Landis and co-workers (Landis 1981, 1983, Landis and Keefe 1983) have shown that the phenotypic transition from noradrenergic to cholinergic demonstrated in culture studies is actually occurring in vivo as part of normal development. The axons innervating the developing sweat glands of rat footpads express the noradrenergic phenotype before they acquire cholinergic properties. Moreover, the time course of this transition is parallel to the appearance

[1] Biology Division, California Institute of Technology, Pasadena, CA 91125, USA

36. Colloquium - Mosbach 1985
Neurobiochemistry

of cholinergic-inducing activity in rat serum (Wolinsky and Patterson 1985). However, it remains to be demonstrated that the transition is caused by the cholinergic factor. In order to determine the role of the cholinergic factor in normal development in vivo, and to understand the mechanism of such neuronal plasticity at the molecular level, therefore, it is essential to purify the factor to homogeneity.

In addition to addressing these basic questions on the development of autonomic neurons, the purified factor can also be used to clarify other interesting issues on the expression of transmitter phenotype. They include: (1) multiple effects of CM. In addition to the alteration in transmitter status, CM changes other characteristics of neurons, such as specific glycoproteins and glycolipids on their surfaces (Chun et al. 1980, Braun et al. 1981, Schwab and Landis 1981, Zurn 1982), and glycoproteins spontaneously released into the medium (Sweadner 1981). (2) regulation of transmitter phenotype in other types of neurons. It has been suggested that the partially purified cholinergic factor can induce ACh synthesis and choline acetyltransferase (CAT) acitivity in both spinal cord (Geiss and Weber 1984) and nodose ganglion neurons (Swerts et al. 1984) as well as sympathetic neurons. (3) relationship to other differentiation factors. Among these are a CAT-stimulating activity of molecular weight (MW) of 40,000-45,000 which affects chick ciliary ganglion neurons in culture (Nishi and Berg 1981), and a CM factor(s) produced by skeletal muscle cells which increases the level of CAT activity of spinal cord neurons in culture (Giller et al. 1977; Godfrey et al. 1980, Brookes et al. 1980, Smith and Appel 1983, Kaufman et al. 1985). (4) effect on the expression of peptidergic phenotypes. It has been suggested that the expression of peptidergic phenotypes is co-regulated with the expression of the noradrenergic/cholinergic phenotype (Kessler 1984, Kessler et al. 1984).

Purification of the factor was started by Michel Weber using serum-containing CM. The partially purified factor has an apparent MW of 40,000-45,000 as determined by Sephadex chromatography. The activity is partially destroyed by treatment with pronase, is extremely sensitive to incubation with periodate, but is unaffected by treatment with urea, β-mercaptoethanol, guanidine-HCl, or incubation with neuraminidase (M.J. Weber 1981).

Since serum in the CM contains a mixture of proteins which must be purified away from the factor, a method of obtaining an active serum-free CM was developed. CM prepared from serum-free medium (without additives) was inactive. However, CM prepared from serum-free medium supplemented with certain hormones based on the results of Sato and co-workers (Bottenstein et al. 1979) was more active than the serum-containing CM. This success in obtaining active serum-free CM not only enabled an enormous purification at the outset but also clarified two other points concerning cholinergic induction. (1) The possibility that the cholinergic factor is a serum component modified by non-neuronal cells was ruled out. (2) Another level of environmental influence on the choice of transmitter phenotype was suggested: namely, that specific hormones can control transmitter choice in developing sympathetic neurons indirectly by affecting the production or release of the factor by non-neuronal cells (Fukada 1980).

The purification from this serum-free, hormone-supplemented conditioned medium has reached at least $10^5$-fold, with a reasonable recovery of activity (Fukada 1983, 1985). The steps include ammonium sulfate precipitation, DEAE- and CM-cellulose column chromotography, Sephadex gel filtration and SDS-polyacylamide gel electrophoresis. In the biological assay used, the unit of activity is expressed as the ACh/CA ratio of cultures which received CM devided by the transmitter ratio of control cultures. The activity precipitates between 60 and 100 % ammonium sulfate. It runs through DEAE cellulose at neutral pH and elutes from CM-cellulose at low salt concentration at neutral pH, indicating that it is a slightly basic molecule. The elution profile from CM-cellulose indicates that there is some charge heterogeneity associated with this molecule, and this point will be discussed later. The two biological activities present in CM, the induction of ACh synthesis and the suppression of CA synthesis, are not separable in any of the purification steps. The activity profile on Sephadex gel filtration appears as a single peak at an apparent MW of 40,000-45,000 which is the same apparent MW determined from serum-containing CM (M.J. Weber 1981). The material from the Sephadex step is very sensitive to treatment with trypsin or proteinase K, indicating that the active factor contains protein. It is, however, resistant to treatment with urea or SDS. Therefore, this Sephadex preparation was further purified by SDS-polyacrylamide gel electrophoresis (SDS-PAGE) under nonreducing conditions (β-mercaptoethanol completely destroys the activity in the presence of SDS). SDS was removed from the gel slices by Dowex ion-exchange columns according to the method of Weber and Kuter (K. Weber and Kuter 1971). A single activity peak was recovered at MW of about 45,000. That the activity has the same apparent MW under denaturing and native conditions suggests that the factor is a single subunit.

The area where activity resides on the SDS-gel is not stained by the silver-staining method of Merril et al. (1981), and is only faintly stained with the modified silver-staining method of Oakley et al. (1980). This area of the gel is also not positive in autoradiograms after iodination with either the Chloramine T, IODO-GEN or lactoperoxidase methods (which label tyrosine residues), but is heavily labeled with the Bolton-Hunter (B-H) reagent (which labels mainly lysine residues under the conditions used) (Bolton-Hunter 1973). The question, then, is: Is the heavily labeled 45 kilodalton (45kd) molecule the cholinergic factor? The findings that the activity and the iodinated 45 kd molecule are both sensitive to trypsin but not chymotrypsin, and that the 45 kd molecule is only labeled by B-H iondination which labels lysine residues support the idea that this labeled 45 kd molecule is the cholinergic factor. In order to further examine this point, the iodinated Sephadex fraction was analyzed by two-dimensional (2-D) gel electrophoresis (O'Farrell et al. 1977); nonequilibrium pH gradient electrophoresis (NEPHGE) was followed by SDS-PAGE. The autoradiogram of this gel shows five discrete spots in the 45 kd region which are spaced at about equal intervals, indicating that the labeled molecule is heterogenous in charge. In order to examine the correlation between these spots and the cholinergic activity, the gel was sliced and the eluted fractions were counted for $^{125}$I-radioactivity and tested for biological activity. The activity is recovered with a high yield and displays five activity peaks. Moreover, the activity peaks and the spaces between the peaks line up precisely with each of the $^{125}$I-labeled protein peaks. However, the sizes of the activity peaks are not proportional to the radioactivity peaks. This could be due to the proteins in the peaks having different degrees of iodination or different degrees of inactivation by iodination. Whatever the reason, the fact that the activity peaks and the $^{125}$I spots line up precisely strongly suggests that the labeled 45 kd molecule is the cholinergic factor.

The charge heterogeneity observed on the 2-D gel warrants comment: Since the activity profile of an unlabeled Sephadex fraction analyzed in the same way also shows charge heterogeneity, the presence of five spots in the iodinated sample is not caused solely by iodination. Often such charge heterogeneity is due to differential glycosylation, particularly sialation (Sparrow et al. 1985). Neuraminidase treatment, however, does not decrease the number of spots, but does shift the spots toward the more basic side of the pH gradient.

Manipulation of the carbohydrate portion of the 45 kd molecule also strongly supports the idea that this protein is the cholinergic factor. When the labeled 45 kd molecule purified from either one-dimensional or two-dimensional gels is treated with endo-β-N-acetylglucosamidase F (Endo F), (an enzyme which cleaves both high-mannose and complex glycoproteins) (Elder and Alexander 1982), six discrete bands result. The lowest MW band is 22 kd, and the activity comigrates with this protein. This result indicates that the cholinergic factor has a core protein of 22 kd, and that the glycosylated factor may migrate anomalously on SDS gels. There are examples of glycoproteins which run anomalously on both Sephadex chromatography and SDS-PAGE (Segrest and Jackson 1972, Felgenhauer 1974, Nozaki et al. 1976). This result is also consistent with the recent report by Weber (Ferrando et al. 1984) that the MW of the cholinergic factor from cultured skeletal muscle cells is 20.6 kd by sedimentation analysis.

The Endo F result also indicates that the bulk of the carbohydrate is not necessary for activity.

In summary, a unique differentiation factor which instructs immature sympathetic neurons to become cholinergic has been purified at least $10^5$-fold in specific activity from serum-free, hormone-supplemented conditioned medium with a reasonable recovery. The most purified fraction is active at less than 10 ng/ml and retains the two biological activities; that is, the induction of ACh synthesis and the suppression of CA synthesis. The factor is a slightly basic glycoprotein whose protein core is composed of a single polypeptide of MW of 22 kd. The biological activity resides in the protein portion of the molecule. N-terminal sequencing of this molecule is currently is progress.

## References


Bolton AE, Hunter WM (1973) Biochem J 133:529-539

Bottenstein J, Hayashi I, Hutchings S, Masui H, Mather J, McClure DB, Ohasa S, Rizzinc A, Sato G, Serrero G, Wolfe R, Wu R (1979) In: Jacoby WB, Pastan IH (eds) Methods in enzymology, vol 45. Academic Press, London New York, pp 94-109

Braun SJ, Sweadner KJ, Patterson PH (1981) J Neurosci 1:1397-1406

Brookes N, Burt DR, Goldberg AM, Bierkamper GG (1980) Brain Res 186:474-479

Bunge R, Johnson M, Ross CD (1978) Science 199:1409-1416

Chun LLY, Patterson PH, Cantor H (1980) J Exp Biol 89:73-83

Claude P (1973) J Cell Biol 59:57a

Elder JH, Alexander S (1982) Proc Natl Acad Sci USA 79:4540-4544

Felgenhauer K (1974) Hoppe-Seyler's Z Physiol Chem 355:1281-1290

Ferrando C, Giess MC, Raynaud B, Swerts JP, Delteil C, Weber MJ (1984) 9th Conf Neurobiol Gif, Aspects Mol Differ Neuron, pp 20-21

Fukada K (1980) Nature (London) 287:553-555

Fukada K (1983) 13th Ann Soc Neurosci Abstr 9(182.5):614

Fukada K (1985) Proc 9th Eur Symp Hormones Cell Regul. Mont Sainte-Odile, pp 377-384

Geiss MC, Weber MJ (1984) J Neurosci 4:1442-1452

Giller EL, Neale JH, Bullock PN, Schrier BK, Nelson PG (1977) J Cell Biol. 74:16-29

Godfrey EW, Schrier BK, Nelson PG (1980) Dev Biol 77:403-418

Hawrot E (1980) Dev Biol 74:136-151

Johnson M, Ross D, Meyers M, Rees R, Bunge R, Wakshull E, Burton H (1976) Nature (London) 262:308-310

Johnson MI, Ross CD, Meyers M, Spitznagel EL, Bunge RP (1980) J Cell Biol 84:680-691

Kaufman LM, Barry SR, Barrett JN (1985) J Neurosci 5:160-166
Kessler JA (1984) Brain Res 321:155-159
Kessler JA, Adler JE, Jonakait GM, Black IB (1984) Dev Biol 103:71-79
Landis SC (1976) Proc Natl Acad Sci USA 73:4220-4224
Landis SC (1980) Devl Biol 77:349-361
Landis SC (1981) In: Garrod D, Feldman JD (eds) Development of the nervous system. Cambridge Univ Press, Cambridge pp 147-160
Landis SC (1983) Fed Proc 49:1633-1638
Landis SC, Keefe D (1983) Dev Biol 98:349-372
Landis SC, MacLeish PR, Potter DD, Furshpan EJ, Patterson PH (1976) 6th Ann Soc Neurosci Abstr 2(280):197
Le Douarin NM (1980) Nature (London) 286:663-669
Le Douarin NM, Teillet M, Ziller C, Smith J (1978) Proc Natl Acad Sci USA 75:2030-2034
Mains RE, Patterson PH (1973a) J Cell Biol 59:329-345
Mains RE, Patterson PH (1973b) J Cell Biol 59:346-360
Mains RE, Patterson PH (1973c) J Cell Biol 59:361-366
Merril CR, Goldman D, Sedman SA, Ebert MH (1981) Science 211:1437-1438
Nishi R, Berg DK (1981) J Neurosci 1:505-513
Nozaki Y, Schechter NM, Reynolds JA, Tanford C (1976) Biochemistry 15:3884-3890
Oakley BR, Kirsch DR, Morris NR (1980) Anal Biochem 105:361-363
O'Farrell PZ, Goodman HM, O'Farrell PH (1977) Cell 12:1133-1142
O'Lague P, Obata K, Claude P, Furshpan EJ, Potter DD (1974) Proc Natl Acad Sci USA 71:3602-3606
Patterson PH (1978) Annu Rev Neurosci 1:1-17
Patterson PH, Chun LLY (1974) Proc Natl Acad Sci USA 71:3607-3610
Patterson PH, Chun LLY (1977a) Dev Biol 56:263-280
Patterson PH, Chun LLY (1977b) Dev Biol 60:473-481
Pintar JE, Maxwell GD, Sweadner KJ, Patterson PH, Breakfield XO (1981) 11th Ann Soc Neurosci Abstr 7(273.9):848
Potter DD, Landis SC, Furshpan EJ (1981) Ciba Found Symp 83:123-138
Rees R, Bunge RP (1974) J Comp Neurol 157:1-11
Reichardt LF, Patterson PH (1977) Nature (London) 270:147-151
Schwab M, Landis S (1981) Dev Biol 84:67-78
Segrest JB, Jackson RL (1972) Methods Enzymol 28B:54-63
Smith RG, Appel SH (1983) Science 219:1079-1081
Sparrow LG, Metcalf D, Hunkapiller MW, Hood LE, Burgess AW (1985) Proc Natl Acad Sci USA 82:292-296
Sweadner K (1981) J Biol Chem 256:4063-4070
Swerts JP, Le Van Thai A, Vigny A, Weber M (1983) Dev Biol 100:1-11
Swerts JP, Giess MC, Mathieu C, Sauron ME, Le Van Thai A, Weber MJ (1984) In: Duprat AM, Kato AC, Weber MJ (eds) The role of cell interaction in early neurogenesis. NATO Adv Inst Ser. Plenum Publ, London, pp 355-344
Weber K, Kuter DJ (1971) J Biol Chem 246:4504-4509
Weber MJ (1981) J Biol Chem 256:3447-3453
Wolinsky E, Patterson PH (1983) J Neurosci 3:1495-1500
Wolinsky EJ, Patterson PH (1985) J Neurosci 5:1509-1512
Zurn AD (1982) Dev Biol 94:483-498

# Biological Properties of Brain-Derived Neurotrophic Factor (BDNF)

Y.-A. Barde[1], R. M. Lindsay[2], and H. Thoenen[1]

## Introduction

An intriguing feature of the development of the vertebrate nervous system is that more neurons are produced than are found in the adult. It is generally thought that the elimination of redundant embryonic neurons serves the purpose of matching the number of neurons of a particular structure with the size of the target to be innervated, as suggested by transplantation and ablation experiments (for review, see Cunningham 1982). The current view is that the target can support only a limited number of neurons and the explanation as to how and why this happens is that the target releases limiting amounts of trophic material that the innervating neurons require at this stage of their development if they are to survive. Considerable support for this view is provided by the work of Levi-Montalcini and Hamburger (1951) and that of Cohen (1960), who discovered and characterized the protein nerve growth factor (NGF) that has all the characteristics of a target-derived trophic molecule required by embryonic neurons at a critical stage of their development, when the axons of these neurons reach their target (for review see Levi-Montalcini and Angeletti 1968, Thoenen and Barde 1980). Particularly relevant in this context are the following results:

1. NGF injected into embryos at the time of neuronal death prevents the death of neurons (Hamburger et al. 1981).
2. Antibodies to NGF greatly enhance the number of dying neurons if administered at the time when those neurons require NGF for survival (Cohen 1960).
3. Target tissues innervated by NGF-responsive neurons contain NGF (Korsching and Thoenen 1983) and NGF messenger RNA (Heumann et al. 1984, Shelton and Reichardt 1984) and both are correlated with the density of innervation.
4. NGF synthesized in the target organ is transported back to the cell body of the innervating neurons and interruption of this transport at the time when neurons depend on NGF for survival leads to the loss of these neurons (for review see Thoenen and Barde 1980).

The fact that NGF is required for survival of two, and only two, types of neuron (the neural crest derived sympathetic and sensory neurons), suggests that there may be a number of such trophic agents. Indeed, many in vitro experiments (of the same kind as those that have led to the successful isolation of NGF) indicate that a variety of tissue extracts or conditioned media contain trophic factors that would support the survival of essentially all the embryonic neurons that have

[1] Max-Planck-Institut für Psychiatrie, Abteilung Neurochemie, D-8033 Planegg-Martinsried, FRG

[2] Sandoz Institute for Medical Research, 5 Gower Place, London WCIE 6 BN, Great Britain

36. Colloquium - Mosbach 1985
Neurobiochemistry

been tested so far, including a large number not responding to NGF (for reviews, see Varon and Adler 1981, Barde et al. 1983, Berg 1984). In the following, we discuss one such trophic factor characterized in our laboratory, brain-derived neurotrophic factor (BDNF).

## BDNF: Purification

A major difficulty encountered in the purification of BDNF is that the specific activity measured in a homogenate of the starting material (pig brain) is very low: approximately 500 μg/ml protein is needed to obtain a half-maximal effect. The effect measured is the number of process-bearing neurons surviving after 2 days in culture, and the source of these neurons is the dorsal root ganglia of the chick at day 10 or 11 of embryonic development. A very large purification factor is thus needed to obtain a protein that appears as one band on SDS-gel electrophoresis, about 1 million-fold (see Barde et al. 1982, for a detailed account of the purification procedure). This is in sharp contrast to the situation with NGF that has allowed so much progress to be made with this protein: indeed, for no apparent reason, NCF is present in very large amounts in the submandibular gland of the adult male mouse, so that a purification factor of only slightly over 100-fold is needed to obtain pure NGF. Both proteins have similar specific activities, on the order of a few ng/ml. Interestingly, the molecular weight of BDNF (12,000) is very similar to that of the monomer of NGF (12,000 to 13,000 depending on how much proteolysis occurs during the purification procedure). Furthermore, both proteins have very basic isoelectric points (9.3 for mouse NGF, 10 or above for BDNF). Comparison of the sequence of BDNF (when available) with that of NGF will decide whether or not both are members of a gene family. In this context, it is interesting to note that in the human genome, only one NGF copy has been detected and that there is no intervening sequence in the region coding for the mature NGF protein (Ullrich et al. 1983).

## BDNF: Biological Properties

Used at a saturating concentration (20 ng/ml), purified BDNF will support the survival and fiber outgrowth of about 30% of the sensory neurons isolated from chick dorsal root ganglia (DRG) at embryonic day 11 (E11). These studies, performed on a polyornithine substrate, were the basis of the assay used for the purification of BDNF. In agreement with result previously obtained with either medium conditioned by C6 glioma cells or rat brain extract (Barde et al. 1980), we found that on such a polycationic substrate, purified BDNF had much smaller effects at early embryonic ages. At E6 for example, BDNF supports only about 5% of the plated neurons. Because modifications of the culture substratum under otherwise identical culture conditions have been shown to markedly enhance neuronal survival (Edgar and Thoenen 1982, Edgar et al. 1984), we decided in a recent study to use the glycoprotein laminin on top of the polyornithine substrate, thinking also that such proteins are more likely to be encountered by developing neurons than a polycationic environment. Using E6 DRG neurons, we found that with laminin about 60% of the neurons survived and put out fibres in the presence of BDNF - about tenfold more than in the absence of laminin. At the same age and under the same conditions, NGF also keeps alive 60% of the plated neurons. The combination of NGF and BDNF keeps alive essentially all the neurons plated (80-100%). These results suggest that the neuronal populations supported in vitro by BDNF or NGF are partially overlapping. The population maintained by NGF (that most likely in vivo gains access to the DRG cell bodies by their peripheral processes, see Korsching and Thoenen 1985) is likely to comprise small, nociceptive, substance P- and somatostatin-containing neurons. Unclear is what the BDNF population corresponds to: radioimmunoassay of neurons grown in the presence of BDNF indicate that some of these neurons might contain substance P (Hayashi and Edgar, unpublished). This could be part of the neuronal population that BDNF and NGF have in common. Another population, specific to BDNF, could consist of larger, proprioceptive neurons, on the basis of the results obtained with BDNF on the purely proprioceptive neurons of the trigeminal mesencephalic nucleus (Davies et al., in preparation). Since BDNF is isolated from the central nervous system (CNS), we speculate that BDNF gains access to the cell bodies of some of the DRG neurons by their central processes. Consistent with this hypothesis is the finding that neurons that have no connexions with CNS tissue, such as ciliary or sympathetic neurons, do not respond to BDNF in culture (Lindsay et al., 1985). On the other hand, we found that the sensory neurons of the nodose ganglion of the chick do respond to BDNF: on a laminin substrate, about 50% of the neurons isolated from E6 chick embryos survive and put out fibres with BDNF (Lindsay et al., 1985). This population of neurons is of particular interest since, in the chick, it has been clearly established that the neurons do not originate from the neural crest, as do DRG neurons, but rather from an epidermal placode. Furthermore, it is now clear that nodose neurons do not respond to NGF and do not have NGF receptors (Davies and Lindsay 1985, Lindsay and Rohrer 1985), results that allow a clear distinction to be made between NGF-responsive and BDNF-responsive sensory neurons.

We are currently examining the possibility that BDNF plays a role in the survival and development of other sensory ganglia, and also the idea that BDNF would be important for the development of neurons associated with specific sensory modalities.

## References


Barde Y-A, Edgar D, Thoenen H (1980) Sensory neurons in culture: changing requirements for survival factors during embryonic development. Proc Natl Acad Sci USA 77:1199-1203

Barde Y-A, Edgar D, Thoenen H (1982) Purification of a new neurotrophic factor from mammalian brain. EMBO J 1:549-553

Barde Y-A, Edgar D, Thoenen H (1983) New neurotrophic factors. Annu Rev Physiol 45:601-612

Berg DK (1984) New neuronal growth factors. Annu Rev Neurosci 7:149-170

Cohen S (1960) Purification of a nerve-growth promoting protein from the mouse salivary gland and its neurocytotoxic antiserum. Proc Natl Acad Sci USA 46:302-311

Cunningham TJ (1982) Naturally occurring neuron death and its regulation by developing neural pathways. Int Rev. Cytol 74:163-186

Davies AM, Lindsay RM (1985) The cranial sensory ganglia in culture: neural crest but not placode-derived neurons respond to nerve growth factor. Dev Biol 3:62-72

Edgar D, Thoenen H (1982) Survival of sympathetic neurons by contact with a conditioned medium factor bound to the culture substrate. Dev Brain Res 5:89-92

Edgar D, Timpl R, Thoenen H (1984) The heparin-binding domain of laminin is responsible for its effects on neurite outgrowth and neuronal survival. EMBO J 3:1463-1468

Hamburger V, Brunso-Bechtold JK, Yip JW (1981) Neuronal death in the spinal ganglia of the chick embryo and its reduction by nerve growth factor. J Neurosci 1:60-71
Heumann R, Korsching S, Scott J, Thoenen H (1984) Relationship between levels of nerve growth factor (NGF) and its messenger RNA in sympathetic ganglia and peripheral target tissues. EMBO J 3:3183-3189
Korsching S, Thoenen H (1983) Nerve growth factor in sympathetic ganglia and corresponding target organs of the rat: correlation with density of sympathetic innervation. Proc Natl Acad Sci USA 80:3513-3516
Korsching S, Thoenen H (1985) Nerve growth factor supply for sensory neurons: site of origin and competition with the sympathetic nervous system. Neurosci Lett, 54:201-205
Levi-Montalcini R, Angeletti PU (1968) Nerve growth factor. Physiol Rev 48:534-569
Levi-Montalcini R, Hamburger V (1951) Selective growth-stimulation effects of mouse sarcoma on the sensory and sympathetic nervous system of the chick embryo. J Exp Zool 116:321-362
Lindsay RM, Rohrer H (1985) Placodal sensory neurons in culture. Nodose ganglion neurons are unresponsive to NGF, lack NGF receptors but are supported by a liver-derived neurotrophic factor. Dev Biol, in press
Lindsay RM, Thoenen H, Barde Y-A (1985) Placode and neural crest-derived sensory neurons are responsive at early developmental stages to brain-derived neurotrophic factor. Dev Biol, in press
Shelton DL, Reichardt LF (1984) Expression of the β-nerve growth factor gene correlates with the density of sympathetic innervation in effector organs. Proc Natl Acad Sci USA 81:7951-7955
Thoenen H, Barde Y-A (1980) Physiology of nerve growth factor. Physiol Rev 60:1284-1335
Ullrich A, Gray A, Berman C, Bull TJ (1983) Human beta-nerve growth factor gene sequence is higly homologous to that of mouse. Nature (London) 303:821-823
Varon S, Adler R (1981) Trophic and specifying factors directed to neuronal cells. In: Fedoroff S, Herz L (ed) Advances in cellular neurobiology, vol II. Academic Press, London New York, pp 115-163

# Multiple Molecular Forms of Acetylcholinesterase and Their Possible Role in the Degradation of Neurohormones

U. Brodbeck[1]

## Introduction

The role of acetylcholinesterase (AChE; EC 3.1.1.7) is well established. It inactivates by hydrolysis acetylcholine, the transmitter substance operating in cholinergic neurotransmission. Besides its activity as an esterase, AChE may also hydrolyze amide bonds contained in aryl-acyl-amides and peptides such as substance P and the enkephalins. AChE occurs either as a soluble enzyme or in association with membranes or the basal lamina (for a review see Massoulié and Bon 1982). In most tissues, several forms of AChE occur side by side and their classification is based on solubility properties and on molecular shape. The most common methods used to distinguish the molecular forms are different extraction procedures (i.e., extractions under various salt conditions, in the absence or presence of a detergent), density gradient centrifugation and gel electrophoresis. Results from such experiments often provide clues to the location of the enzyme in vivo.

AChE contained in the extracellular matrix is attached to the basal lamina from which it is solubilized in high ionic strength conditions. This form of AChE is highly asymmetric, as four catalytic subunits are attached via disulfide bridges to a collagen-like tail that interacts with the proteins of the basal lamina, thereby attaching AChE to the extracellular matrix. AChE solubilized from the basal lamina exists in multiple molecular forms in which two or three basic units form collagen-like threads and contain 8 or 12 catalytical subunits. Depending on the number of catalytic subunits contained in the oligomeric assembly, these asymmetric forms are denominated $A_4$, $A_8$, and $A_{12}$-AChE. Electric organs and skeletal muscle of the electric eel *(Electrophorus electricus)* are exceptional, as they contain almost exclusively collagen-tailed AChE, mostly as the $A_{12}$ form. In *Torpedo* electric organ only abou 50% of AChE comprises asymmetric, collagen-tailed forms. Viratelle and Bernhard (1980) showed that the remaining activity could be solubilized only in the presence of micellar amounts of a detergent. They purified that form of AChE in the presence of 1% sodium deoxycholate and showed that the pure enzyme had similar catalytic properties as the high salt-soluble one, but it lacked the collagen tail. From their results these authors concluded that in *Torpedo* electric organ AChE is localized in situ in the phospholipid membrane as well as in the basal lamina. Independently, Bon and Massoulié (1980) described the total solubilization of AChE from electric organs of *Torpedo marmorata* in high salt buffer containing 1% Triton X-100. They discerned two different subpopulations of AChE, i.e. the collagen-tailed, asymmetric forms that did not bind micellar amounts of Triton X-100 and a detergent soluble AChE which formed polydisperse aggregates in absence of the detergent and which by treatment with pronase or proteinase K lost its detergent-binding

[1]Medizinisch-chemisches Institut der Universität Bern, Bühlstrasse 28, CH-3012 Bern, Switzerland

36. Colloquium - Mosbach 1985
Neurobiochemistry

property. From these and numerous other studies (cf. Massoulié and Bon 1982), it became clear that AChE functioning in cholinergic neurotransmission is localized in the extracellular matrix, where it is either attached via its collagen-like tail to the basal lamina or presumably anchored into the lipid bilayer of the plasma membranes.

Besides its occurrence in *Torpedo* electric organ, membrane-bound, detergent-soluble AChE has been identified in brain, muscle, and erythrocytes, and is defined as that enzyme activity that remains in a $6 \times 10^6$ g × min supernatant solution only in the presence of micellar amounts of detergents. It is that part of AChE which is amphiphilic, i.e., the molecules contain both a hydrophilic and a hydrophobic domain in their structure. Detergent-soluble AChE is designated as DS-AChE or form D-AChE and the number of catalytic subunits is given by a subscript number ($D_1$, $D_2$, $D_4$ etc.)

Besides forms A- and D- of AChE, there exists yet a third one that neither contains a collagen-like tail nor is membrane-bound. It is a globular protein soluble in buffer of physiological ionic strength; it is denominated as form G-AChE or low salt-soluble AChE. This form either represents the intermediary product in the biosynthesis of the asymmetric forms before attachment of the catalytic subunit to the collagen-like tail, or, as suggested by Chubb et al. (1976), may be a secretory form of the enzyme, that in brain might have a function other than hydrolyzing acetylcholine. Within the scope of the present review, the occurrence, structural and molecular properties of membrane-bound detergent-soluble AChE are discussed and a special emphasis is given to the enzyme occurring in brain and the possible role of the secretory form of AChE.

## Purification and Catalytic Properties of Detergent-Soluble AChE

DS-AChE has been purified from a variety of neuronal and non-neuronal sources. These include the electric organ of different *Torpedo* species (Viratelle and Bernhard 1980, Lee et al. 1982, Stieger and Brodbeck 1985), bovine brain (Ruess et al. 1976), chicken brain (Rotundo 1984), human brain (Sørensen et al. 1982, Gennari and Brodbeck 1985) and rat brain (Rakonczay et al. 1981). Erythrocyte membranes are a convenient starting material to isolate DS-AChE, although the function of the enzyme in this cell remains obscure (Niday et al. 1977, Ott et al. 1975, 1982, Grossmann and Liefländer 1979).

In all purification procedures the DS-AChE was present as protein-detergent mixed micelle. When the enzyme was kept in the neutral detergent Triton X-100 throughout the purification, the specific activities of the purified enzyme ranged between 3100 IU/mg or protein (Rakonczay et al. 1981) and 5700 IU/mg of protein (Ott et al. 1982). On the other hand, Viratelle and Bernhard (1980) solubilized and purified the enzyme in the presence of the negatively charged detergent sodium deoxycholate and reported a specific activity of 200 IU/mg of pure protein. Niday et al. (1977) purified the enzyme in Triton X-100, followed by a step in which sodium deoxycholate was used and the specific activity of their pure product was 418 IU/mg of protein. In both cases SDS polyacrylamide gel electrophoresis revealed a single band only, indicating that despite the much lower specific activity the enzyme did not contain visible amounts of contaminating protein.

A possible explanation for the large difference in specific activity of a seemingly pure protein from the same source (i.e., 418 vs 5700 IU/mg of DS-AChE from human red cell membrane) could be that the activity of DS-AChE is sensitive to the surrounding amphiphilic environment. First evidence for this notion was obtained by Sihotang (1976). He observed that the activity of DS-AChE from human erythrocytes decreased after solubilization in sodium deoxycholate with decreasing phospholipid content of the extract and that readdition of membrane lipids stimulated enzyme activity upto 3.5-fold. Hall and Brodbeck (1978) noted that pure DS-AChE from human erythrocytes became rapidly and irreversibly inactivated in the presence of micellar amounts of sodium deoxycholate, the rate of inactivation being dependent on the concentration of detergent used. Inclusion of lipid in the incubation mixture abolished the inactivation at low sodium deoxycholate concentration and reduced it at higher concentrations of the detergent. Reconstitution into liposomes by dialyzing the enzyme in a solution containing sodium deoxycholate and lecithin preserved enzyme activity. Wiedmer et al. (1979) further investigated the interaction of different detergents with pure DS-AChE from human erythrocytes, and showed that the activity of the enzyme depends on hydrophobic interactions. When the enzyme was deprived of the stabilizing detergent, a rapid, time-dependent decay of enzyme activity with a half life of about 10 s was seen. Inactivation was stopped by readdition of detergent, but once the enzyme was fully inactive, readdition of any amphiphilic molecule could not restore enzyme activity.

The dependence of enzyme activity on the presence of stabilizing amphiphiles is also observed with DS-AChE from *Torpedo marmorata* (Stieger and Brodbeck 1985), human brain (Sørensen et al. 1982) and from bovine brain (Landauer et al. 1982). Although the requirements for hydrophobic interactions appear to be common to DS-AChE from different sources, there are marked differences in their detergent sensitivity. This is exemplified by the stability of the enzyme in β-D-octylglucoside. DS-AChE from human red cell membranes is initially less active in β-D-octylglucoside than in Triton X-100 and in addition slowly loses activity. The enzyme from human brain is even less stable and that from *Torpedo marmorata* extremely unstable in β-D-octyl-glucoside (Brodbeck and Ott 1984).

It could be further demonstrated that phospholipids also sustained full enzyme activity. This was shown by diluting DS-AChE into solution containing increasing amounts of phospholipid liposomes. At low lipid concentrations, virtually no enzyme activity was obtained, but it increased with increasing lipid concentrations. This lipid dependency so far has been observed with DS-AChE from human eryhtrocytes (Wiedmer et al. 1979), Di Francesco and Brodbeck 1981), with the enzyme from human caudate nucleus (Sørensen et al. 1982) and with that from *Torpedo marmorata* (Stieger, unpublished). As discussed later, removal of detergent in suitable conditions leads to the formation of protein micelles consisting of aggregated DS-AChE. These protein micelles were soluble and active in buffer devoid of detergent. The aggregated enzyme may be regarded as an amphiphile indpendent entity. In the case of DS-AChE from human red cell membranes, the range in protein concentration in which aggregation occurs could be assessed by measuring the amphiphile-independent activity of (aggregated) DS-AChE. At protein concentrations below 10 ng/ml, the enzyme was present in its protomeric, amphiphile dependent state, whereas above 2.5 μg/ml the enzyme was fully aggregated and its activity was amphiphile-independent (Wiedmer et al. 1979).

## Molecular Properties of DS-AChE

These forms of DS-AChE that have been investigated so far either consist of 2 or 4 subunits. DS-AChE from human red cell membranes and from the electric organ of *Torpedo* is dimeric ($D_2$-AChE) whereas the majority of DS-AChE in brain is tetrameric ($D_4$-AChE). The molecular weight of the enzyme from human red cell membranes has been measured by analytical ultracentrifugation and values of 151,000 ± 8000 were obtained (Ott et al. 1982). In this thermodynamic method, octyltetraoxyethylene was used as solubilizing detergent, which permits molecular weight determinations without measuring the amount of detergent bound (Rosenbusch et al. 1982). Rosenberry and Scoggin (1984) calculated a molecular weight of 160,000 ± 8000 from the hydrodynamic properties of the enzyme and from the amount of Triton bound to it. Molecular weights were reported for native DS-AChE from bovine brain (340,000, Ruess et al. 1976, Ruess and Liefländer 1981, Landauer et al. 1984); chicken brain (460,000, Rotundo 1984); human brain (300,000 Sørensen et al. 1982) and rat brain (318,000, Rakonczay et al. 1981). These numbers should be viewed, however, with caution, as they represent either apparent molecular weights assessed by electrophoresis in nondenaturing conditions or were calculated from the Stokes radii without correction for bound detergent.

The subunits in all $D_2$-forms appear to be identical and active in catalysis. Results obtained by Römer-Lüthi et al. (1980) and by Ott et al. (1983) suggest that the red cell enzyme is a dimer in the native membrane as well as in reconstituted phospholipid vesicles. On the other hand, DS-AChE from brain is isolated as a tetrameric species. According to Landauer et al. (1984), the $D_4$ enzyme from bovine brain contains only two active sites per tetramer, and they suggested that this form consists of pairs of dislike subunits. SDS-polyacrylamide gel electrophoresis in nonreducing conditions of form $D_4$ from human brain also showed two pairs of subunits appearing at 150,000 and 130,000 daltons (Gennari and Brodbeck 1985). The functional molarity of this enzyme, however, was determined as four active sites per tetramer (Gentinetta and Brodbeck, unpublished).

The molecular weights obtained in nonreducing conditions are in all forms of DS-AChE about twice that obtained in reducing conditions, which suggests that pairs of subunits are interlinked by disulfide bridges. Ott et al. (1983) converted pure $D_2$-AChE from human red cell membranes to a stable $D_1$-form by treating the enzyme with 2-mercaptoethanol and iodoacetic acid. The recovery of enzyme activity was more than 60%. When eryhtrocyte ghosts were similarly treated, the enzyme could subsequently be solubilized as monomeric species and more than 90% of activity was recovered. A similar observation has been made with AChE contained in a high salt detergent extract of bovine superior cervical ganglion (Vigny et al. 1979). On the other hand, a similar treatment of DS-AChE from *Torpedo marmorata* or from human brain has so far failed to produce monomeric enzyme (Gennari and Stieger, unpublished).

Recently, Rosenberry and Scoggin (1984) showed that the monomerization of $D_2$-AChE from human red cell membranes produced 1.72 to 1.86 SH-groups per dimer, and from these numbers it was concluded that pairs of subunits are interlinked by a single disulfide bridge.

## Amphiphilic Nature of DS-AChE

It is now well established that DS-AChE is an amphiphilic protein that possesses distinct hydrophilic and hydrophobic domains. The former protrudes into the aqueous phase and contains the active center, whereas the latter anchors the enzyme into the lipid bilayer of the membrane. Disruption of the membrane by detergent is a prerequisite to solubilize amphiphilic membrane proteins. Such proteins stay in solution by detergent binding and will aggregate if the solubilizing agent is removed. The binding of detergents to proteins may be demonstrated by sucrose density gradient centrifugation or charge shift-crossed immunoelectrophoresis. The latter method relies on the observation that amphiphilic proteins show changes in their electrophoretic mobility when they bind detergent micelles differing in charge (Bhakdi et al. 1977, Helenius and Simons 1977). This is achieved by electrophoresis of proteins in a first dimension on agarose gels containing (a) uncharged micelles of Triton X-100, (b) negatively charged micelles composed of sodium deoxycholate and Triton X-100 and (c) positively charged micelles consisting of cetyl-trimethylammonium bromide and Triton X-100 (for experimental details see Bjerrum 1981, Stieger and Brodbeck 1984).

Charge shift-crossed immunoelectrophoresis has been used to verify the binding of detergents to DS-AChE from human erythrocytes (Bhakdi et al. 1977, Weitz et al. 1984), human caudate nucleus (Sørensen et al. 1982) and *Torpedo marmorata* (Stieger and Brodbeck 1985). Results obtained with this method showed that DS-AChE has a lower anodic mobility in negatively charged micelles and a higher one in positively charged micelles as compared to the mobility in an uncharged micelle. Using [$^3$H]Triton X-100, Rosenberry and Scoggin (1984) determined the number of Triton X-100 bound to DS-AChE from human red cell membranes and found that 0.63 g Triton X-100 was bound per g of $D_2$-AChE and 0.75 g/g of $D_1$-AChE. These ratios corresponded to aobut 140 molecules of Triton X-100 bound to the dimer and approximately 80 bound to the monomer. As a homologous micelle of Triton X-100 contains about 120 molecules (Robinson and Tanford 1975), the values of 140 and 80 indicate that each subunit individually interacts with a detergent micelle. This observation provides further evidence that each subunit of the dimer contains a hydrophobic domain which must be sufficiently apart from each other to allow the individual binding of a 45,000 dalton micelle (Rosenberry and Scoggin 1984).

## Aggregation of DS-AChE in the Absence of Detergents

As mentioned above, amphiphilic proteins aggregate to protein micelles upon removal of the solubilizing detergent. The resulting aggregates are often polydisperse and become water-soluble by mutually covering the hydrophobic domain in the protein (Helenius and Simon 1975). This is also observed with DS-AChE from human red cell membranes (Ott et al. 1975, Ott and Brodbeck 1978) with the enzyme from human brain (Sørensen et al. 1982, Gennari and Brodbeck 1985), bovine brain (Ruess and Liefländer 1981), *Torpedo californica* (Lee et al. 1982) and *Torpedo marmorata* (Stieger and Brodbeck 1985). In the aggregated state, enzyme activity no longer depends on the presence of detergent molecules. Here hydrophobic interactions among individual enzyme molecules lead to the formation of protein micelles in which the enzyme is kept in a catalytically active conformation.

## How is DS-AChE Anchored to Membranes?

Available experimental evidence supports the notion that DS-AChE is anchored into the lipid bilayer of biological and artificial membranes through a short hydrophobic peptide that may be labeled by the photo-activatable reagent 3-trifluoromethyl-3-(m-$^{125}$I-iodophenyl)diazirine (Stieger et al. 1984). This reagent has been used to specifically label the hydrophobic domain of various membrane proteins (Brunner 1981, Brunner and Semenza 1981). It partitions nearly quantitatively into the lipid phase of membranes and liposomes, into detergent and protein-detergent mixed micelles. Upon photoactivation a carbene is generated, which covalently reacts with molecules in the hydrophobic phase. In the native *Torpedo* enzyme, the radioactivity was associated with the poly-peptide chain that upon SDS polyacrylamide gel electrophoresis migrated with an apparent molecular weight of 67,000. Limited digestion with proteinase K lead to a hydrophilic form devoid of radioactivity that showed an apparent molecular weight of 66,000 and to a labeled peptide which appeared at about 3000 daltons and presumably contained the membrane-binding domain. The hydrophobic peptide was isolated by chromatography on Sephadex LH 60 in ethanol/formic acid, and by gel filtration its molecular weight was determined at 3100, which was in close agreement to that estimated by SDS polyacrylamide gel electrophoresis (Stieger et al. 1984). Limited proteolysis thus removed a hydrophobic entity which binds detergents and is essential for reconstitution (Stieger and Brodbeck 1985). These findings parallel those obtained by Rosenberry et al. (1984) and Dutta-Choudhury and Rosenberry (1984) with DS-AChE from human erythrocytes. Limited digestion of that enzyme with papain also quantitatively released the hydrophobic $^{125}$I-TID label. These authors further showed that the hydrophobic domain corresponds to the C-terminus of the polypeptide chain.

An alternative way of attachment of DS-AChE to biological membranes through a specific interaction involving phosphatidylinositol (PI) has been proposed. DS-AChE from erythrocytes could apparently be released from membranes without the use of detergent by a PI-specific phospholipase from *Staphylococcus aureus* (Low and Finean 1977) and from *Bacillus thuringiensis* (Ikezawa et al. 1983, Taguchi et al. 1984). PI-specific phospholipase C also released DS-AChE from *Torpedo* electric tissue (Futerman et al. 1983, 1985) and the solubilized enzyme behaved like a hydrophilic form G. From their results these authors concluded that DS-AChE is associated with the plasma membrane of *Torpedo* electric organ through a strong and specific interaction which involves PI, either by a covalent or a noncovalent linkage.

## Does AChE in Brain Have Functions Other than Hydrolyzing Acetylcholine?

The idea that AChE may have a function other than the hydrolysis of acetylcholine was advanced by Silver (1974), who based her idea on the observation that in certain regions of the brain AChE is present in high levels while there is little or no choline acetyltransferase (ChAT) and acetylcholine present. These regions include the substantia nigra, cerebellum, globus pallidus and the hypothalamus. Using immunochemical methods Levey et al. (1983) demonstrated that in the hypothalamus and substantia nigra groups of individual neurons contained AChE but were devoid of ChAT. Presently it is impossible to list all neuronal groups were AChE exists beyond its function in cholinergic transmission and the fact that AChE is present either alone or in excessive amounts

Table 1. Effect of bovine (brain) acetylcholinesterase on various peptides (Chubb et al. 1983)

| | | |
|---|---|---|
| Peptides that are hydrolyzed by AChE | | |
| Enkephalins | | Products |
| [leu$^5$] enkephalin (tyr, gly, gly, phe, leu) | | tyr + leu |
| [met$^5$] enkephalin (tyr, gly, gly, phe, met) | | tyr + met |
| Other peptides | | |
| leu, arg | tyr, gly | |
| phe, leu | gly, gly, gly | |
| leu, met | Substance P | |
| Peptides that are not hydrolzyed by AChE | | |
| gly, gly | β-endorphin | |
| pro, gly, gly | angiotensin | |
| BSA | oxytocin | |
| bombesin | vasopressin | |

over ChAT does not prove that AChE has noncholinergic functions in the brain. There is indirect evidence that suggests that within the caudate nucleus AChE is present in both somatostatin and GABA containing neurons (Bolam et al. 1984). AChE is also found in close association with substance P containing cells in the retina (Salipan et al. 1983) and the dorsal horn of the spinal cord (Chubb et al. 1980).

As mentioned earlier, AChE in brain also exists as soluble enzyme, termed by Chubb the secretory form of AChE (Chubb et al. 1976). It is neither collagen-tailed nor membrane-bound and has a higher rate of synthesis than the membrane-bound form. What could be the function of this soluble form? The idea that the enzyme might have a function in the extracellular space arose from observations that the enzyme is present in the cerebrospinal fluid due to a physiological release process from neurons (Chubb 1984, Greenfield 1984). As pointed out by the latter, levels of AChE in cerebrospinal fluid do not correspond to the release of acetylcholine, as:

1. In brain acetylcholine is released mainly into the lateral ventricle whereas the highest AChE-content is found in the cisterna magna.
2. The release of acetylcholine into the ventricular system is depressed by anesthesia but the release of AChE is not (Greenfield 1980).
3. If certain brain areas such as the caudate nucleus or the substantia nigra are stimulated electrically, there is an increase of AChE in cisternal cerebrospinal fluid that is not blocked by cholinergic receptor antagonists (Greenfield and Smith 1979).

Much of AChE in cerebrospinal fluid originates from nigrastriatal neurons but only the dopamine-containing neurons release the enzyme. There is no ChAT in and no chalinergic input to the neurons, which raises the questions on the substrate for AChE secreted from these neurons.

It is well established that AChE, besides acting as an esterase, is also able to catalyze the hydrolysis of amide bonds. Fujimoto (1974) reported the existence in brain of an enzyme capable of cleaving o-nitroacetanilide to o-nitroaniline and acetate. From further work of Fujimoto (1976) and that of Balasubramanian, it became clear that the aryl-acyl-amidase activity was exhibited by AChE (Oommen and Balasubramanian 1977, 1978, George and Balasubramanian 1980). The ary-acyl-amidase activity of AChE is not only sensitive to inhibition of acetylcholine and its analogs, but also by serotonin and tryptamine, whereas it is insensitive towards dopamine, histamine, and tyramine (for a short review cf. Balasubramanian 1984). Results obtained from chemical modification of AChE allowed Balasubramanian to propose a model accounting for both aryl-acylamidase as well as esterase activity of AChE (Majumdar and Balasubramanian 1984).

Chubb et al. (1983) addressed the question directly to the hydrolysis of naturally occurring as well as synthetic model peptides and found that among a number of different compounds the enkephalins and substance P were hydrolyzed by AChE (Table 1). It should be noted, however, that the turnover number for peptide substrates are orders of magnitudes lower than for the natural substrate acetylcholine. In the study of Chubb et al. (1983), the enkephalins only served as model peptides and these authors did not imply that these compounds are the "natural peptide substrates", so to date their real nature remains unknown. Soluble AChE occurs both intra- and extracellularly. Thus AChE located inside neuronal cells might play a part in the intracellular processing of the precursors of neuropeptides. This notion was tested by Millar and Chubb (1984), who demonstrated that the treatment of chick retina tissue sections with purified AChE increased the immunoreactivity of enkephalins and substance P. From their results, Miller and Chubb concluded that the peptidase activity of AChE has the capacity to hydrolyze proteins of which some may be the precursor molecules for the enkephalins and substance P. In addition that part of AChE which is secreted into the extracellular space might be involved in the degradation of neuropeptides.

## References


Balasubramanian AS (1984) Have cholinesterases more than one function? TINS Dec 84: 467-468

Bhakdi S, Bhakdi-Lehnen B, Bjerrum OJ (1977) Detection of amphiphilic proteins and peptides in complex mixtures. Charge-shift crossed immunoelectrophoresis and two-dimensional charge-shift electrophoresis. Biochim Biophys Acta 470:35-44

Bjerrumg OJ (1981) Analysis of membrane antigens by means of qualitative detergent immunoelectrophoresis. In: Azzi A, Brodbeck U, Zahler P (eds) Membrane proteins, a laboratory manual. Springer, Berlin Heidelberg New York, pp 13-42

Bolam IP, Ingham CA, Smith AD (1984) The section-Golgi-impregnation procedure 3. Combination of Golgi-impregnation with enzyme histochemistry and electron microscopy to characterize acetylcholinesterase containing neurons in the rat neostriatum. Neuroscience 12:687-709

Bon S, Massoulié J (1980) Collagen-tailed and hydrophobic components of acetylcholinesterase in torpedo marmorata electric organ. Proc Natl Acad Sci USA 77:4464-4468

Brodbeck U, Ott P (1984) Amphiphile dependency and amphiphilic structure of detergent soluble acetylcholinesterase. In: Brzin M, Barnard EA, Sket D (eds) Cholinesterases: Fundamental and applied aspects. De Gruyter, Berlin, pp 187-201

Brunner J (1981) Labeling of the hydrophobic core of membranes. Trends Biochem Sci 6:44-46

Brunner J, Semenza G (1981) Selective labeling of the hydrophobic core of membranes with 3-trifluoromethyl-3-(m$^{125}$I-iodophenyl) diazirine, a carbene generating reagent. Biochemistry 20:7174-7182

Chubb IW (1984) Acetylcholinesterase: multiple functions. In: Brzin M, Barnard EA, Sket D (eds) Cholinesterases: Fundamental and applied aspects. De Gruyter, Berlin, pp345-359

Chubb IW, Goodman S, Smith AD (1976) Is acetylcholinesterase secreted from central neurons into the cerebrospinal fluid? Neuroscience 1:57-62
Chubb IW, Hodgson AJ, White GH (1980) Acetylcholinesterase hydrolyzes substance P. Neuroscience 5:2065-2072
Chubb IW, Ranieri E, White GH, Hodgson AJ (1983) The enkephalins are amongst the peptides hydrolyzed by purified acetylcholinesterase. Neurscience 10:1369-1377
Di Francesco C, Brodbeck U (1981) Interaction of human red cell membrane acetylcholinesterase with phospholipids. Biochim Biophys Acta 640:359-364
Dutta-Choudhury TA, Rosenberry TL (1984) Human erythrocyte acetylcholinesterase is an amphiphatic protein whose short membrane-binding domain is removed by papain digestion. J Biol Chem 259:5653-5660
Fujimoto D (1974) Serotonin-sensitive ary-acylamidase in rat brain. Biochem Biophys Res Commun 61:72-74
Fujimoto D (1976) Serotonin-sensitive ary-acylamidase activity of acetylcholinesterase. FEBS Lett 71:121-123
Futerman AH, Low MG, Silman I (1983) A hydrophobic dimer of actylcholinersterase from torpedo california electric organ is solubilized by phosphatidylinositol-specific phospholipase C. Neursci Lett 40:85-89
Futerman AH, Fiorini RM, Roth E, Low MG, Silman I (1985) Physiochemical behavior and structural characteristics of membrane-bound acetylcholinesterase from torpedo electric organ - Effect of phosphatidylinositol-specific phospholipase C. Biochem J 226:369-377
Gennari K, Brodbeck U (1985) Molecular forms of acetylcholinesterase from human caudate nucleus: comparison of salt soluble and detergent soluble tetrameric enzyme species. J Neurochem 44:697-704
George ST, Balasubramanian AS (1980) The identity of the serotonin-sensitive arylacylamidase with acetylcholinesterase from human erythrocytes, sheep basal ganglia and electric eel. Eur J Biochem 111:511-524
Greenfield SA (1980) Acetylcholinesterase in cerebrospinal fluid. In: Brzin M, Sket D, Bachelard H (eds) Synaptic constituents in health and disease. Pergamon Press, Oxford, pp 438-439
Greefield SA (1984) A novel function for acetylcholinesterase in nigrostriatal neurons. In: Brzin M, Barnard EA, Sket D (eds) Cholinesterase: Fundamental and applied aspects. De Gruyter, Berlin, pp 289-303
Greenfield SA, Smith AD (1979) The influence of electric stimulation of certain brain regions on the concentration of acetylcholinesterase in rabbit cerebrospinal fluid. Brain Res 177:445-459
Grossmann H, Liefländer M (1979) Acetylcholinesterase from bovine erythrocytes. Purification and properties of the enzyme solubilized in the presence and the absence of Triton X-100. Z Naturforsch 34c:721-725
Hall R, Brodbeck U (1978) Human erythrocyte membrane acetylcholinesterase. Incorporation into the lipid bilayer structure of liposomes. Eur J Biochem 89:159-167
Helenius A, Simons K (1975) Solubilization of membranes by detergents. Biochim Biophys Acta 415:29-79
Helenius A, Simons K (1977) Charge-shift electrophoresis. Simple method for distinguishing between amphiphilic and hydrophilic proteins. Proc Natl Acad Sci USA 74:529-532
Ikezawa H, Nakabayashi T, Suzuki K, Nakaijama M, Taguchi T, Taguchi R (1983) Complete purification of phosphatidylinositol-specific phospholipase C from a strain of bacillus thuringiensis. J Biochem 93:1717-1719
Landauer P, Ruess KP, Liefländer M (1982) Modulation of acetylcholinesterase activity by glycoside-detergents and their solubilization efficiency for neuronal membranes from bovine nucleus caudatus. Biochem Biophys Res Commun 106:848-855
Landauer P, Ruess KP, Liefländer M (1984) Bovine nucleus caudatus acetylcholinesterase: Active site determination and investigation of a dimeric form obtained by selective proteolysis. J Neurochem 43:799-805
Lee SL, Heinemann S, Taylor P (1982) Structural characterization of the asymmetric (17 + 13) S forms of acetylcholinesterase from torpedo. J Biol Chem 257:12283-12291
Levey AI, Wainer BM, Mufson EJ, Mesulam MM (1983) Co-localization of acetylcholinesterase and choline acetyltransferase in the rat cerebrum. Neuroscience 9:9-22

Low MG, Finean JB (1977) Non-lytic release of acetylcholinesterase from erythrocytes by a phosphatidylinositol-specific phospholipase C. FEBS Lett 82:143-146
Majumdar R, Balasubramanian AS (1984) Chemical modification of acetylcholinesterase from eel and basal ganglia: effect on the acetylcholinesterase and aryl-acylamidase activities. Biochemistry 23:4088-4093
Massoulié J, Bon S (1982) The molecular forms of cholinesterase and acetylcholinesterase in vertebrates. Annu Rev Neursci 5:57-106
Millar TJ, Chubb IW (1984) Treatment of sections of chick retina with acetylcholinesterase increases the enkephalin and substance P immuno-reactivity. Neuroscience 12:441-451
Niday E, Wang CS, Alaupovic P (1977) Studies on the characterization of human erythrocyte acetylcholinesterase and its interaction with antibodies. Biochim Biophys Acta 469:180-193
Oommen A, Balasubramanian AS (1977) The inhibition of brain aryl-acylamidase by 5-hydroxytryptamine and acetylcholine. Biochem Pharmacol 26:2163-2167
Oommon A, Balasubramanian AS (1978) Aryl-acylamidase of monkey brain and liver: response to inhibitors and relationship to acetylcholinesterase. Biochem Pharmacol 27:891-895
Ott P, Brodbeck U (1978) Multiple molecular forms of acetylcholinesterase from human erythrocyte membranes. Eur J Biochem 88:119-125
Ott P, Jenny B, Brodbeck U (1975) Multiple molecular forms of purified human erythrocyte acetylcholinesterase. Eur J Biochem 57:469-480
Ott P, Lustig A, Brodbeck U, Rosenbusch JP (1982) Acetylcholinesterase from human erythrocyte membranes: dimers as functional units. FEBS Lett 138:187-189
Ott P, Ariano BH, Bringgeli Y, Brodbeck U (1983) A monomeric form of human erythrocyte membrane acetylcholinesterase. Biochim Biophys Acta 729:193-199
Rakonczay Z, Mallol J, Schenk H, Vincendon G, Zanetta JP (1981) Purification and properties of the membrane-bound acetylcholinesterase from adult rat brain. Biochim Biophys Acta 657:243-256
Robinson NC, Tanford C (1975) The binding of deoxycholate, Triton X-100, sodium dodecyl sulfate and phosphatidylcholine vesicles to cytochrome b5. Biochemistry 14:369-378
Römer-Lüthi CR, Ott P, Brodbeck U (1980) Reconstitution of human erythrocyte membrane acetylcholinesterase in phospholipid vesicles. Analysis of the molecular forms by cross-linking studies. Biochim Biophys Acta 601:123-133
Rosenberry TL, Scoggin DM (1984) Structure of human erythrocyte acetylcholinesterase. J Biol Chem 259:5643-5652
Rosenberry TL, Scoggin DM, Dutta-Choudhury TA, Haas R (1984) Human erythrocyte acetylcholinesterase is an amphiphatic form. In: Brzin M, Barnard EA, Sket D (eds) Cholinesterases: Fundamental and applied aspects. De Gruyter, Berlin, pp 155-172
Rosenbusch JP, Garavito RM, Dorset DL, Engel A (1982) Structure and Function of a pore-forming transmembrane protein: high resolution studies of a bacterial porin. In: Peters H (ed) Protides of the biological fluids. Colloq 29. Pergamon Press, Oxford, pp 171-174
Rotundo RL (1984) Purification and properties of the membrane-bound form of acetylcholinesterase from chicken brain. J Biol Chem 259:13189-13194
Ruess KP, Liefländer M (1981) Molecular forms of purified cytoplasmatic and membrane bound bovine-brain-acetylcholinesterase solubilized by different methods. Z Naturforsch 36c:968-972
Ruess KP, Weinert M, Liefländer M (1976) Affinitätschromatographische Reinigung der Acetylcholinesterase aus Nucleus Caudatus vom Rind im Grossmassstab. Hoppe-Seyler's Z Physiol Chem 358:1543-1550
Salipan TJ, Morgan IG, Chubb IW (1983) Decreased concentrations of acetylcholinesterase and leu-enkephalin in dark-adapted chick retinae. Neurosci Lett Suppl II S73
Sihotang K (1976) Acetylcholinesterase and its association with lipid. Eur J Biochem 63:519-524
Silver A (1974) The biology of cholinesterases. In: Neuberger A, Tatum EL (eds) Frontiers of biology, vol 36. Elsevier/North Holland, Amsterdam
Sørensen K, Gentinetta R, Brodbeck U (1982) An amphiphile-dependent form of human brain caudate nucleus acetylcholinesterase: purification and properties. J Neurochem 39:1050-1060

Stieger S, Brodbeck U (1984) Investigation of amphiphilic nature of different forms of acetylcholinesterase from torpedo marmorata by charge shift crossed immunoelectrophoresis. In: Azzi A, Brodbeck U, Zahler P (eds) Enzyme, receptors and carriers of biological membranes. Springer, Berlin Heidelberg New York, pp 13-19

Stieger S, Brodbeck U (1985) Amphiphilic detergent soluble acetylcholinesterase from torpedo marmorata: Characterization and conversion by proteolysis to a hydrophilic form. J Neurochem 44:48-56

Stieger S, Brodbeck U, Reber B, Brunner J (1984) Hydrophobic labeling of the membrane binding domain of acetylcholinesterase from torpedo marmorata. FEBS Lett 168:231-234

Taguchi R, Suzuki K, Nakabayashi T, Ikezawa H (1984) Acetylcholinesterase release from mamalian erythrocytes by phosphatidylinositol-specific phospholipase C of bacillus thuringiensis and characterization of the released enzyme. J Biochem 96: 437-446

Vigny M, Bon S, Massoulié J, Gisiger V (1979) The subunit structure of mamalian acetylcholinesterase: catalytic subunits, dissociating effects of protelysis and disulfide reduction on the polymeric forms. J Neurochem 33:559-565

Viratelle OM, Bernhard SA (1980) Major component of acetylcholinesterase in torpedo electroplax is not basal lamina associated. Biochemistry 19:4999-5007

Weitz M, Bjerrum OJ, Brodbeck U (1984) Characterization of an active hydrophilic erythrocyte membrane acetylcholinesterase obtained by limited proteolysis of the purified enzyme. Biochim Biophys Acta 776:65-74

Wiedmer T, Di Francesco C, Brodbeck U (1979) Effects of amphiphiles on structure and activity of human erythrocyte membrane acetylcholinesterase. Eur J Biochem 102: 59-64

# Synthesis of Neurohypophyseal Hormones

D. Richter[1], R. Ivell[1], H. Schmale[1], P. Nahke[1], and B. Krisch[2]

## Introduction

Oligopeptides which function as hormones, neurotransmitters, or paracrine modulators are generally synthesized as longer precursors which may comprise other biologically active peptides and which, in analogy to the similarly constructed viral polyproteins, have been termed cellular polyproteins (Fig. 1, p. 34; Koch and Richter 1980, Herbert 1981).

Conversion of the polyproteins into the mature peptides is a crucial step in their biosynthetic pathways and requires precise signals as well as specific enzymes(s) recognizing them. The signals generally comprise pairs of basic amino acids flanking the respective peptides.

## Vasopressin and Oxytocin: Precursor and Gene Structures

The nonapeptide vasopressin is known to control water resorption in the distal kidney tubuli, oxytocin milk ejection, and the contraction of the uterus during birth. In addition, the two hormones appear to be involved in many other processes, such as cardiovascular regulation, control of body temperature, or brain development to name a few (De Wied 1983).

The precursors to the hormones vasopressin and oxytocin are typical examples of cellular polyproteins. Both are synthesized in the magnocellular neurons of the hypothalamus and transported axonally to the posterior pituitary. Besides the hormones the composite precursors contain the functionally linked carrier proteins, called neurophysins. Only the vasopressin precursor includes a third moiety, a glycoprotein at its C-terminus.

So far the oxytocin and vasopressin genes have been determined from rat (Schmale et al. 1983, Ivell and Richter 1984) and cow (Ruppert et al. 1984). Each gene comprises three exons separated by two intervening sequences (Fig. 2, p. 35). Each exon encodes one of the principal functional domains of the polyprotein – hormone, carrier protein, glycoprotein; in the case of the oxytocin gene, the third exon comprises only the C terminal variable part of the neurophysin plus an extra basic amino acid.

The structural organization of the vasopressin and oxytocin genes is in line with other eukaryotic split genes and includes the consensus

[1]Institut für Zellbiochemie und klinische Neurobiologie, Universität Hamburg, D-2000 Hamburg 20, FRG

[2]Anatomisches Institut der Universität Kiel, D-2300 Kiel, FRG

36. Colloquium - Mosbach 1985
Neurobiochemistry

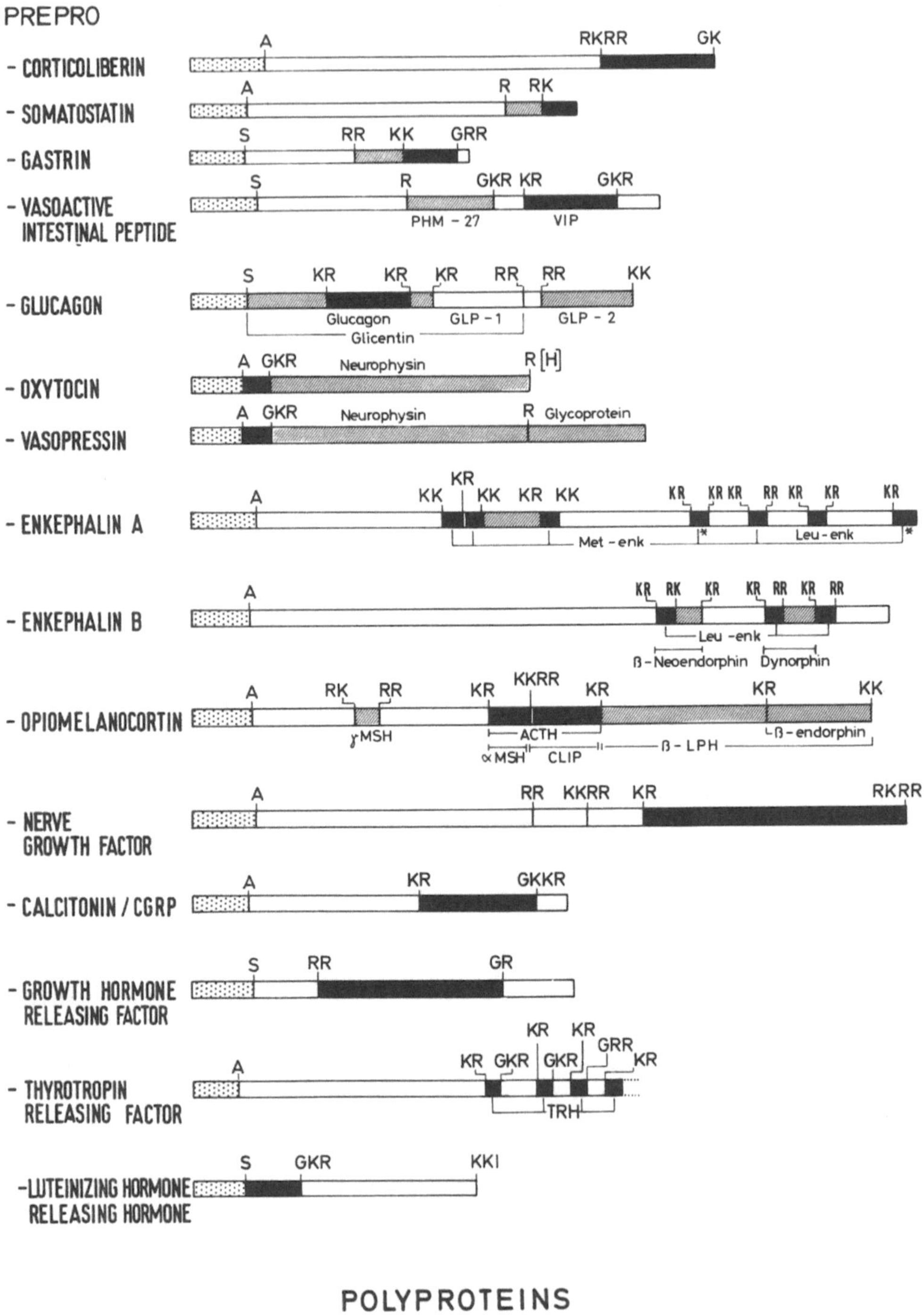
PREPRO
- CORTICOLIBERIN
A
RKRR
GK
- SOMATOSTATIN
A
R RK
- GASTRIN
S
RR KK GRR
- VASOACTIVE INTESTINAL PEPTIDE
S
R
GKR KR
GKR
PHM - 27
VIP
- GLUCAGON
S KR KR KR RR RR KK
Glucagon
GLP - 1
GLP - 2
Glicentin
- OXYTOCIN
A GKR
Neurophysin
R [H]
- VASOPRESSIN
A GKR
Neurophysin
R Glycoprotein
- ENKEPHALIN A
A
KK KR KK KR KK
KR KR KR RR KR KR KR
Met - enk
Leu - enk
- ENKEPHALIN B
A
KR RK KR KR RR KR RR
Leu -enk
ß - Neoendorphin
Dynorphin
- OPIOMELANOCORTIN
A
RK RR
KR KKRR KR
KR KK
γMSH
ACTH
αMSH
CLIP
ß - LPH
ß - endorphin
- NERVE GROWTH FACTOR
A
RR KKRR KR
RKRR
- CALCITONIN / CGRP
A
KR
GKKR
- GROWTH HORMONE RELEASING FACTOR
S
RR
GR
- THYROTROPIN RELEASING FACTOR
A
KR GKR GKR
KR KR GRR KR
TRH
-LUTEINIZING HORMONE RELEASING HORMONE
S GKR
KKI
POLYPROTEINS

Fig. 1

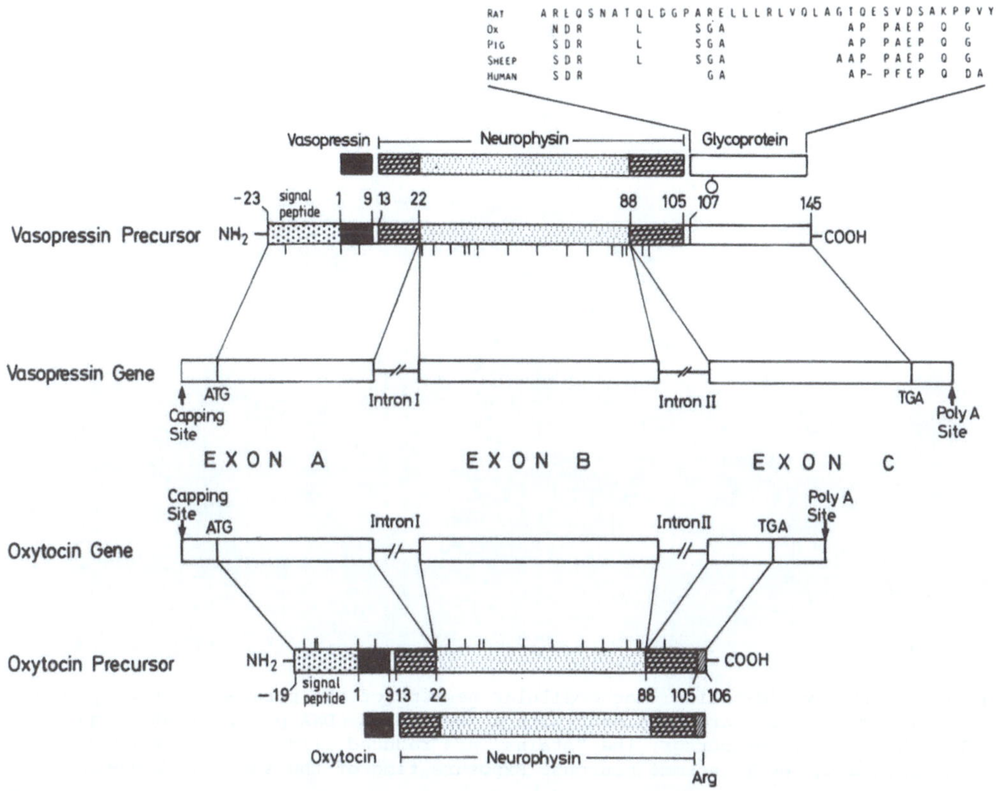


Fig. 2. Comparision of the vasopressin and oxytocin genes from rats

sequences at exon-intron junctions, a modified Goldberg-Hogness sequence, CATAAAT, located 29 nucleotides upstream of the presumptive transcription start site, an A residue bounded by a pyrimidine-rich region at the transcription start site, and a polyadenylation site AATAAA at the 3' end of the sequence.

←

Fig. 1. Polyproteins as precursors to neuropeptides. *Dotted bars* signal sequence; *open bars* "spacer" sequences (bars are not drawn to scale); *black bars* the physiologically active hormones or neuropeptides; *hatched bars* auxiliary sequences known to modify the physiology of the associated peptide hormone; *ACTH* adrenocorticoprotein; *CGRP* calcitonin gene-related peptide; *CLIP* corticotropin-like intermediate-lobe peptide; *GLP-1*, *GLP-2* glucagon-like peptide 1,2; *b-LPH* b-lipotropin; *PHM-27* a 27-amino acid peptide with an N-terminal histidine and an C-terminal methionine; *Met** C-terminally extended methionine enkephalin; *MSH* melanotropin-stimulating hormone; *TRH* thyrotropin-releasing factor; *VIP* vasoactive intestinal peptide; amino acids occurring as processing signals or as extension of peptides are listed as follows: *A* alanine; *G* glycine; *H* histidine; *K* lysine; *I* isoleucine; *R* arginine; *S* serine

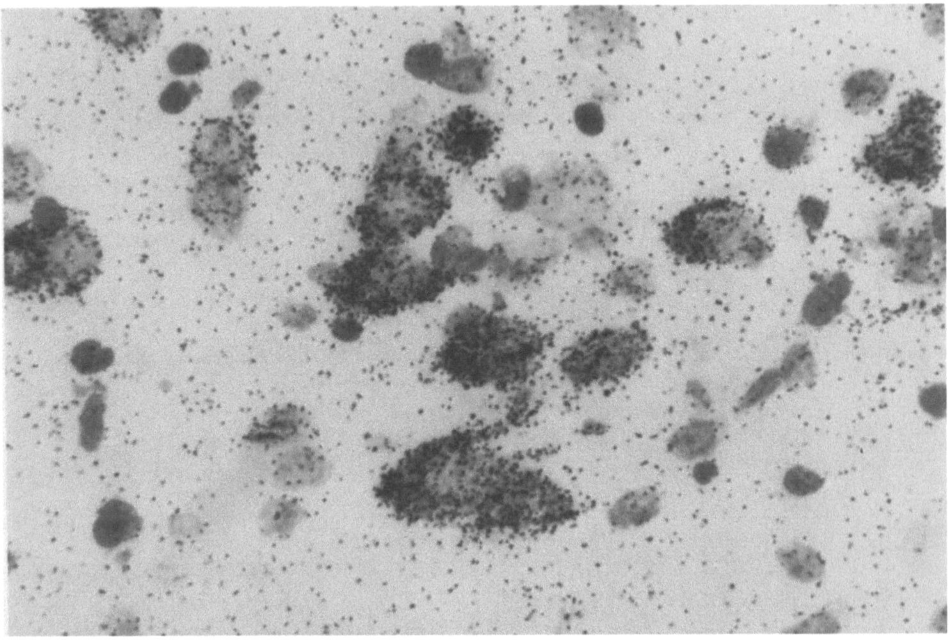

Fig. 3. In situ hybridization: magnocellular neurons of the paraventricular nucleus of the rat hypothalamus are "stained" with a $^{3}H$-labeled cDNA probe encoding specifically the oxytocin precursor. The "stains" are reduced silver grains and reflect the oxytocin-mRNA level in some neurons. Exposure time of the section: 5 weeks

When comparing the protein-coding region of the two genes from rat and cow, the homologies in the sequence for the second exon are particularly striking. The vasopressin and oxytocin gene from rat are entirely homologous within exon B for a stretch of 143 nucleotides with one base difference. The homology is even more striking for the respective bovine sequences, extending even into the preceding intron covering 332 bp without any interruption. A recent gene conversion event may explain this rather unusually high degree of sequence homology.

## Gene Regulation

With DNA probes specific for either hormone precursor, it is now possible to study their regulation at the transcriptional level. Two methods have been adopted to quantify the mRNA encoding the vasopressin precursor; one is based on liquid or blot hybridization assays, the other makes use of thin layer sections of the hypothalamus, which are then hybridized in situ to specific radiolabeled DNA probes.

In the following, some examples are given for elevated mRNA levels under different physiological conditions. Rats placed under osmotic stress by drinking 2% saline respond with an elevated plasma osmolality. To maintain the water balance the stores of vasopressin in the posterior pituitary are depleted, while the hormone level in the plasma rises (Robertson 1977). In this situation there is a significant increase of the mRNA encoding vasopressin (Burbach et al. 1984). Interestingly, the increase is greatest in the nucleus supraopticus, less in the nucleus paraventricularis, and absent in the suprachiasmatic nucleus. It is known that vasopressin-producing neurons of the SON and PVN project towards the posterior pituitary, those of the SCN towards other brain areas. Hence, it has been speculated that these data point to a specific regulation in the expression of the vasopressin gene in different hypothalamic regions (Burbach et al. 1984).

Whether the different levels of vasopressin-encoding mRNA in the three hypothalamic areas assayed reflect indeed differential regulation of the vasopressin gene has to be examined more rigorously, for instance by transcription run-off experiments measuring the levels of nascent vasopressin-mRNA in the cell nucleus. Also other physiological parameters are needed which may affect specifically the vasopressin mRNA level in the SCN but not in the other two areas.

Another method to test levels of mRNA is in situ hybridization. This method combines a number of advantages; it allows not only the cells expressing the respective gene to be identified, but also the mRNA in these cells to be quantified, as well as a study of the effect of endogenously induced factors on the anatomy and number of cells expressing the respective gene.

The expression of the rat oxytocin gene in the SON is demonstrated in Figure 3 (in collaboration with J. Morrell, J. McCabe, D. Pfaff, The Rockefeller University, New York; manuscript in preparation). It shows that some neuronal cell bodies are specifically decorated with silver grains whereas others are not, suggesting that the former are oxytocin-producing cells. When oxytocin mRNA levels are assayed in female rats and compared to 15-day pregnant rats the grain numbers increased significantly in the pregnant rats (in collaboration with J. McCabe, J. Morrell, D. Pfaff, manuscript in preparation). These data are in good agreement with those showing that during pregnancy the oxytocin level increases in the plasma and the pituitary (Boer et al. 1979). Although preliminary, these few examples serve to illustrate the considerable potential of the in situ hybridization method.

## Conversion of the Precursors

A still unresolved problem is the processing of the precursors and the order of the processing steps which lead to the mature peptides. In general, when precursor synthesis has been accomplished in the rough endoplasmic reticulum, the protein is transferred to the Golgi apparatus. There it becomes concentrated and formation of neurosecretory granules begins with the precursor undergoing a number of maturation steps. In the case of the vasopressin and oxytocin precursors there is ample time for these steps during their transport along the axons of the neuronal cells. Early kinetic studies (Sachs and Takabatake 1964) indicate that in the dog this transport requires about 1.5 h from the time of synthesis to release of the physiological nonapeptide.

Little information is yet available on the sequence of maturation for the vasopressin or oxytocin precursors. A glycopeptide has been isolated by Concanavalin A Sepharose chromatography from neurosecretory granules, which contains a neurophysin but not a vasopressin antigenic moiety (North et al. 1983). A similar intermediate has been detected in the guinea pig posterior pituitary (Robinson and Jones 1984), implying that in the neurohypophysis the first processing step occurs between vasopressin and the neurophysin.

In a heterologous system such as the oocytes of *Xenopus laevis*, which have been programmed with hypothalamic mRNA by microinjection, the

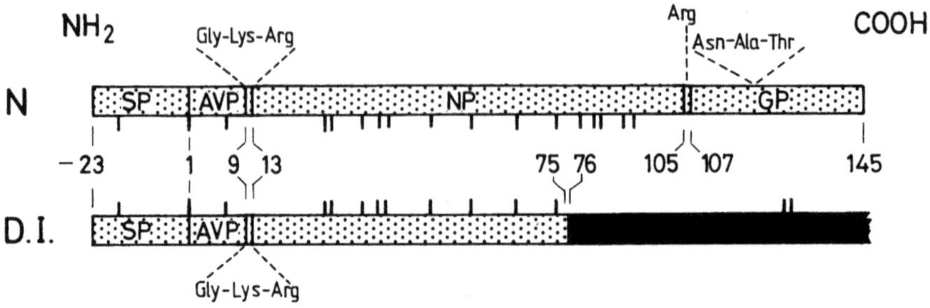


Fig. 4. Comparison of the vasopressin precursors from wild-tye *(N)* and diabetes insipidus *(D.I.)* rats deduced from their gene structures. (For details see Schmale and Richter 1984)

cleavage of the vasopressin precursor appears to occur between the neurophysin and the glycoprotein at the single arginine residue (Richter 1983). The second cleavage step between hormone and neurophysin has not been observed in this system. This would imply that the processing enzymes of the oocyte recognize a single arginine, but not a pair of basic amino acids, suggesting that there are at least two different processing systems in the cell, one acting on pairs and the other on single basic amino acids.

*Signals and Processing Enzymes.* A number of signals can be recognized within the vasopressin and oxytocin precursors that serve either for the proteolytic cleavage process (single or pairs of basic amino acids), for glycosylation (Asn-X-Thr) or for amidation (Gly). The significance of these signals for the processing event is stressed by the finding that the nucleotide sequence encoding glycine, lysine, and arginine – the three amino acids that separate the hormones from their respective carrier neurophysins – are highly conserved between the two genes.

In recent years a number of processing enzymes with trypsin- or carboxypeptidase B-like activity have been described (Steiner et al. 1983). So far these studies lack either a detailed analysis of the processed product(s) or quite often the polyproteins to be studied are replaced by model peptides. Basically, two groups of processing enzymes have been proposed as candidates for catalyzing the conversion of polyproteins into mature peptides: *serine proteases*, including kallikreins and plasmin or plasminogen activator and *thiol proteases* with cathepsin B-like properties. It is generally assumed that these processing enzymes contain a broad substrate specificity, e.g., when cDNA encoding human proinsulin is expressed in a tumor cell line normally secreting ACTH, processing of the proinsulin precursor into insulin occurs correctly (Moore et al. 1983).

## Genetic Defect – Diabetes Insipidus

Lack of vasopressin in vivo leads to the excretion of large amounts of a very dilute urine accompanied by excessive thirst. This disorder is known as vasopressin-sensitive diabetes insipidus, in contrast to the nephrogenic form which is due to vasopressin resistance of the kidney (Sokol and Valtin 1982).

The defect can be reversed by applying arginine vasopressin. In man, X-linked and autosomally dominant forms of diabetes insipidus exist (Green et al. 1967). An appropriate model for studying the genetic defect is the so-called Brattleboro rat with an autosomally recessive form of diabetes insipidus (Sokol and Valtin 1982). These animals lack vasopressin as well as its neurophysin, while oxytocin and its associated carrier are not affected.

The gene for the vasopressin precursor from the mutant rats has been isolated and its sequence determined (Schmale and Richter 1984, Schmale et al. 1984). It contains a deletion of a single G residue in exon B which encodes the conserved part of neurophysin. This mutation gives rise to an open reading frame predicting a hormone precursor with a different C terminus (Fig. 4). Because of the open reading frame there is no stop codon for terminating the translation process. Hence theoretically translation of the mRNA could lead to a product with a polylysine tail at the C terminus. The new reading frame of the mutated gene would also predict the replacement of a basic amino acid (arginine) normally separating the neurophysin from the glycoprotein, as well as the loss of the glycosylation site.

## Expression of the Mutated Vasopressin Gene

*Transcription and mRNA Sequence.* A number of experiments have demonstrated that the mutated gene is correctly transcribed and that the resulting mRNA contains a single nucleotide deletion in the neurophysin-encoding part, as predicted from the gene sequence (Schmale and Richter 1984, Schmale et al. 1984).

Inspection of Northern blots of hypothalamic poly(A)$^+$ RNA from wild-type (Wistar) and Brattleboro rats suggests comparable amounts of vasopressin and oxytocin precursor-specific mRNA (Schmale and Richter 1984). Quantitative studies by dot blot analysis of hypothalamic poly(A)$^+$ RNA from Long-Evans, Wistar, and homozygous Brattleboro rats indicate the following relative amounts of vasopressin precursor-specific mRNA: Long-Evans 100%, Wister 66%, and Brattleboro 27%. Information from cloning experiments suggests that the vasopressin precursor-specific mRNA amounts to 0.04% of total poly(A)$^+$ RNA in the Brattleboro, to 0.1% in the Wister and 0.15% in the Long-Evans rat.

*Translation.* Though vasopressin-encoding mRNA can be isolated from hypothalami of Brattleboro rats, the question arises whether this mRNA is translatable. Since the deletion of a nucleotide residue in the second half of the mRNA sequence leads to a shift in the reading frame in a way that a stop codon signal is no longer read, the precise termination of the nascent peptide chain should not occur. In theory, the ribosome should read through the 3'-end of the mRNA including the poly(A) sequence giving rise to a larger precursor product. If one calculates the maximal size of such a potential precursor than roughly 70 amino acids have to be added to the normal-sized protein (MW 19,000) giving rise to a molecular weight between 26,000 and 28,000. The C terminus of this precursor should consist of ca. 50 lysine residues corresponding to the 150 adenosines of the poly(A) tail.

Indeed when Brattleboro hypothalamic mRNA is translated in a cell-free system the normal-sized vasopressin precursor is not observed, instead small amounts of vasopressin-like precursors have been identified with molecular weights in the range from 19,000 to 26,000 (Ivell, unpublished). The heterogeneously sized translation products may reflect how far these ribosomes have been able to translate the mRNA. The prod-

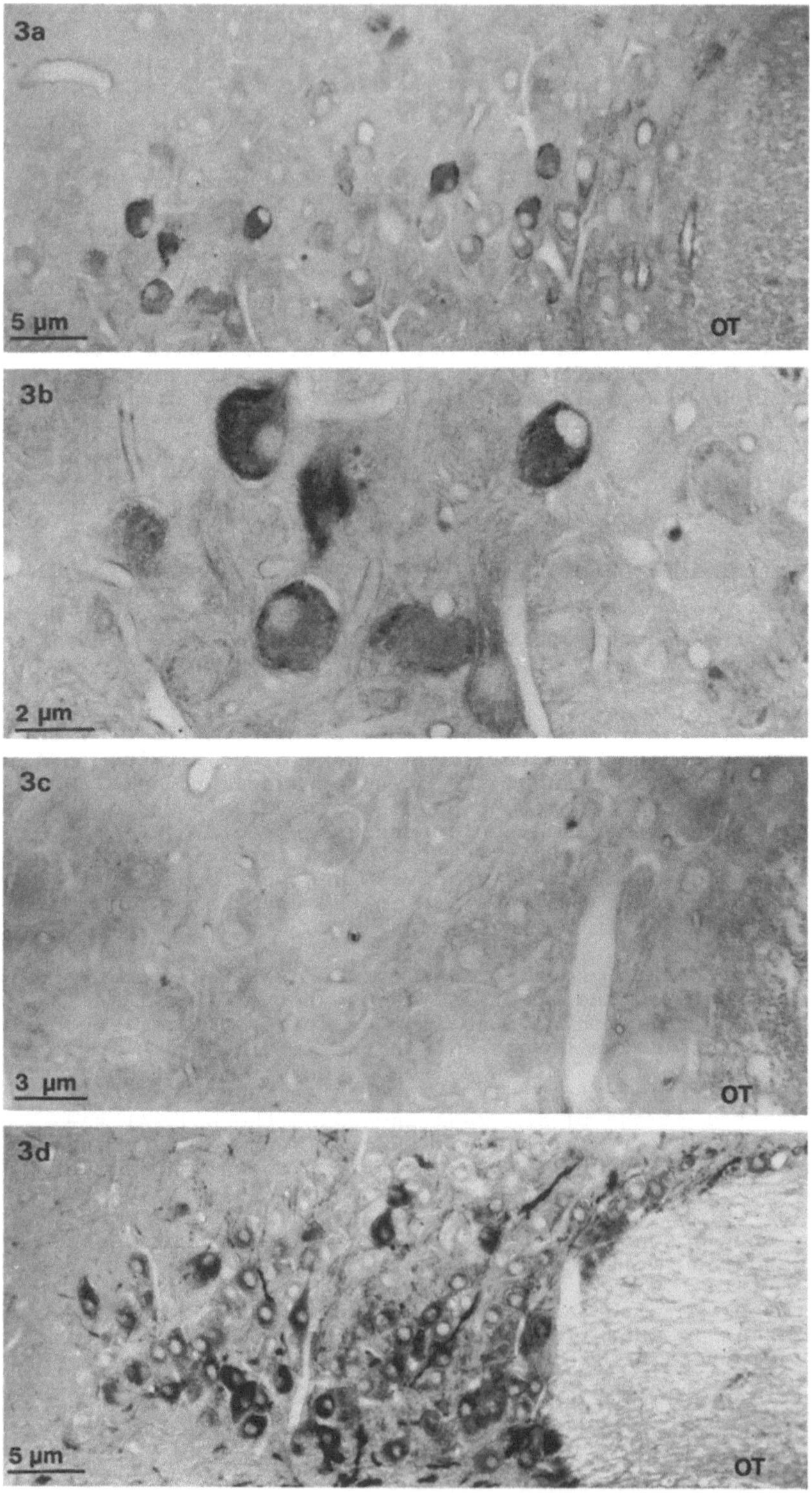


Fig. 5

ucts have been identified by using antisera raised either against vasopressin, neurophysin or a 14mer peptide predicted from the new reading frame and specific for the mutated Brattleboro vasopressin precursor (Schmale et al. 1984).

*Immunocytochemical Studies.* Using antibodies against the 14mer peptide (C-peptide antibodies) in immunocytochemical studies at the light microscopic level, it can be demonstrated that the enlarged cells of the supraoptic nucleus can be specifically stained in Brattleboro, but not in wild-type rat hypothalami (Fig. 5). Typically the stain is located in the vicinity of the Nissl bodies. Preliminary electron microscopic studies indicate that this stain is situated in the rough endoplasmic reticulum, in the inner cisternae of the Golgi apparatus and in small lysosomes. Since the mutated vasopressin precursor is highly charged at its C terminus (ca. 50 lysine residues!), it is not too surprising that most of the synthesized material is fixed at the membrane. Staining is restricted to the cell body and cannot be found in axons or in the pituitary. Very few cells are stained in the paraventricular nucleus, none in the suprachiasmatic nucleus.

*Conclusion.* The Brattleboro rat represents the rare case where a single nucleotide deletion leads to an open reading frame predicting a protein precursor which in theory should end in a poly lysine tail at its C terminus. Another case with a frameshift mutation has been reported for hemoglobin Wayne which is an alpha chain variant and which has been explained by assuming a minus one frameshift. This leads to the extension of the hemoglobin chain by five amino acids to be followed then by a stop codon (Seid-Akhavan et al. 1976). In that example only a few nucleotide residues of the 3' untranslated region and none of the poly(A) sequence are copied into protein.

Though the vasopressin gene is expressed in the hypothalamus of Brattleboro rats, relatively few cells in the supraoptic nucleus can be stained immunocytochemically. This could be due to the modified precursor structure which because of its highly charged C-terminus might be folded in a way not easily accessible to the antibodies used. The reduced vasopressin precursor synthesis, however, may be limited just because of the poly(A) sequence exhausting the supply of lysine, its tRNA or the corresponding aminoacyl-tRNA synthetase.

Alternatively, translation of mRNA and packaging of the synthesized product into neurosecretory granules may be a more coordinated and feed-back controlled process than generally expected. The electron microscopic studies so far show that most of the mutated vasopressin precursor is found in the rough endoplasmic reticulum, in terminal Golgi vesicles and in lysosomes indicating that the "false" precursor is recognized and discarded in an unknown way by the cell. Is the modified C-peptide the wrong tag which does not allow its correct packaging in the neurosecretory granules? Is the formation of the correct precursor and neurosecretory granules in a closely linked process which is halted when certain signals within the protein structure are absent?

◄

Fig. 5. Immunohistochemical staining of the supraoptic nucleus of hypothalami from Brattleboro (a-c) and normal rats (d). Staining procecudure: peroxidase-anti peroxidase method using either antisera raised against the C-peptide of the mutated vasopressin precursor (a-c) or against vasopressin (d). In c the antibodies were inactivated by preincubation with the antigen. a x 220; b x 560; c x 340; d x 220; *OT* optic tract

Or is the mutated precursor simply discarded because of its abnormal poly-lysine structure at its C-terminus? No doubt there are many questions still left to be answered before we shall completely understand the genetic defect in diabetes insipidus rats.

## References


Boer K, Dogterom J, Snijdewint FJ (1979) Endocrinology 80:41P-42P

Braverman LE, Mancini JP, McGoldrick DM (1965) Ann Intern Med 63:503-508

Burbach JPH, De Hoop MJ, Schmale H, Richter D, De Kloet FR, Ten Haaf JA, De Wied DJ (1984) Neuroendocrinology 39:582-584

De Wied D (1983) Progr Brain Res 60:155-167

Douglass J, Civelli O, Herbert E (1984) Annu Rev Biochem 53:665-715

Green JR, Buchan GC, Alvord EC jr, Swanson AG (1967) Brain 90:707-714

Herbert E (1981) Trends Biochem Sci 6:184-188

Ivell R, Richter D (1984) Proc Natl Acad Sci USA 81:2006-2010

Koch G, Richter D (eds) (1980) Biosynthesis, modification, and processing of cellular and viral polyproteins. Academic Press, London New York

Moore HPH, Walker MD, Lee F, Kelly RB (1983) Cell 35:531-538

North WG, Mitchell TI, North GM (1983) FEBS Lett 152:29-34

Richter D (1983) Trends Biochem 8:278-281

Robertson GI (1977) Recent Prog Horm Res 33:333-385

Robinson ICAF, Jones PM (1984) Neurosci Lett 39:273-278

Ruppert D, Scherer G, Schutz G (1984) Nature (London) 308:554-557

Russell JT, Brownstein MJ, Gainer H (1980) Endocrinology 107:1880-1891

Sachs H, Takabatake Y (1964) Endocrinology 75:943-948

Schmale H, Richter D (1984) Nature 308:705-709

Schmale H, Heinsohn S, Richter D (1983) EMBO J 2:763-767

Schmale H, Ivell R, Breindl M, Darmer D, Richter D (1984) EMBO J 3:3289-3293

Seid-Akhavan M, Winter WP, Abramson RK, Rucknagel DL (1976) Proc Natl Acad Sci USA 73:882-886

Sokol HW, Valtin H (eds) (1982) The Brattleboro rat. Annu NY Acad Sci 394:1-828

Steiner DF, Docherty K, Chan SJ, Segundo BS, Carroll R (1983) in: Koch G, Richter D (eds) Biochemical and clinical aspects of neuropeptides. Academic Press, London New York, pp 3-13

# Degradation and Biological Inactivation of Neuropeptides

K. Bauer[1]

## Introduction

An ever-increasing number of neuropeptides have been identified in recent years. These substances act as neurohormones, neuromodulators, neurotransmitters or paracrine effector substances on a variety of target cells via binding to specific receptors. As for other signal-transmitting substances, it is mandatory that highly efficient degradation and/or elimination systems also exist for neuropeptides. Otherwise, the concentration of these substances would build up so that further secretion would lead only to a marginal increase in their concentration, making their function as messengers impossible. This is especially true for neuronal communication factors. Of course, for substances which are released with high frequency, these mechanisms not only have to be very efficient but they have to be also very rapid.

In vitro, rapid degradation by blood and tissue enzymes has been observed for all neuropeptides, but the physiological functions of these enzymes within the mechanisms of neuropeptide metabolism remained so far obscure. Principally, we should anticipate that these enzymes might serve different functions at different levels.

For the neuropeptide-synthesizing cells, some as yet unknown mechanisms must exist to regulate intracellular concentrations of these peptides. For example, under conditions where secretion is suppressed, these mechanisms might act as a security device system. However, since the site of synthesis, storage, and release necessarily have to be separated from the site of degradation, the neuropeptide-degrading enzymes cannot act directly on these peptides and, therefore, these enzymes cannot fulfill a regulatory function. The mechanisms which determine the translocation of the peptides (such as the mechanisms leading to the crinophagic disposal of secretion granules etc.) apparently represent the *limiting* and, thus, regulatory control elements which finally determine the biological inactivation of the neuropeptides. These mechanisms are as yet unknown, and experimentally most difficult to evaluate, since hydrolysis of the neuropeptides does not lead to specific metabolites but only the cell constituent amino acids. In selected cases, the availability of enzyme-specific inhibitors might provide the tools to explore these mechanisms and also to obtain some information concerning the yet unknown turnover rates of neuropeptides.

After release of neuropeptides and their interaction with the receptors, some mechanisms must exist to clear the target site for the transmission of the next signal.

---

[1]Institut für Biochemie und Molekulare Biologie, TU Berlin, Franklinstr. 29, D-1000 Berlin 10

36. Colloquium - Mosbach 1985
Neurobiochemistry

As shown for some neuropeptides and several proteohormones, peptides may enter the target cell by receptor-mediated endocytosis. The fate of the peptide-receptor complex is not known in most cases, but conceivably the peptides are degraded by lysosomal or cytosolic enzymes. By nature, the internalization of peptides is a relatively slow process, but may represent an adequate mechanism for peptides which are released with very low frequency and also in cases where the peptides exhibit an extremely low $K_D$.

For other peptides these processes are obviously too slow. After termination of action (which occurs exclusively at the receptor level), for example through dissociation of the peptide-receptor complex, the peptide could be removed from the site of target interaction by diffusion. At the pituitary level, the peptide might subsequently be simply washed out and enter the peripheral circulation. Thereby, the peptide would already be diluted to biologically ineffective concentrations. Subsequently, the peptide could be degraded by enzymes present, e.g., in the major organs of metabolic clearance, the liver and the kidneys (metabolic clearance function of peripheral neuropeptide-degrading enzymes). In the CNS, the dissociated peptide could diffuse out of the synaptic cleft, but should be degraded thereafter by enzymes which may be associated with other cells (e.g., glial cells, endothelial cells, etc.) or localized on neurons remote from the receptor sites. Obviously, such enzymes which serve general scavenger functions are biologically most important: they maintain the system of directed communication by preventing the peptides from diffusion to other target sites and their accumulation in the CSF. Correspondingly, these enzymes could also influence the results obtained from pharmacological studies by preventing exogeneously applied peptides from reaching the target sites.

It is striking, for example, that some peptides, such as the enkephalins, do not elicit any effect even when injected intraventricularly. Only when these peptides are administered in the presence of appropriate peptidase inhibitors can some effects be observed.

For peptides which are released with high frequency, the synaptic cleft should be cleared very rapidly for the transmission of the next signal. In this case, simple diffusion processes alone may be too slow and we might expect that the removal of peptidergic transmitters from the synaptic cleft is facilitated by additional mechanisms as known, e.g., for classical neurotransmitters.

Catecholamines, for example, are known to be rapidly removed from the synaptic cleft by re-uptake into the secreting cell, where they are metabolized by intracellular enzymes. For neuropeptides, however, inactivation through energy-dependent uptake process has not yet been convincingly demonstrated and there is little evidence that such mechanisms exist.

By analogy with the hydrolysis of acetylcholine by acetylcholinesterase, it seems more likely that peptidergic neurotransmitters are degraded by peptidases localized in the vicinity of the receptors. It is clear that such enzymes cannot act as "switch" for turning off peptidergic signals, but can only fulfill a specialized scavenger function. For this "target site clearance function" we should expect that these enzymes exhibit a high turnover number and that they are present in sufficiently high concentrations. Such enzymes, however, do not necessarily have to have a high affinity and specificity for a given peptide. Primarily determined by their specific localization, even peptidases with limited substrate specificity could directly influence the transmission of the peptidergic signal. In this context it should be noted that from a biochemical point of view acetylcholinesterase is also not an absolutely specific enzyme. The biological specific function of acetylcholinesterase is perhaps mainly due to the fact that acetylcholine is the only biological important representative of this class of substances. In contrast, neuropeptides belong to a very complex family and structurally they are also much more complex. Moreover, the degradation of neuropeptides is catalyzed not only by one, but by several enzymes, and the different fragments formed are subject to further degradation by other enzymes. The degradation of these substances to the free amino acids as final degradation products gives rise to a very complex fragmentation pattern. In addition, there are no biochemical criteria to distinguish peptidases serving target site clearance functions from enzymes serving other functions. For these reasons, it is not surprising that a "peptidergic acetylcholinesterase" has not yet been identified.

To find an understanding on the yet unknown biological functions of neuropeptide-degrading enzymes, it is prerequisitory first to delineate the pathways of neuropeptide fragmentation and to evaluate the biochemical properties of the individual enzymes capable of hydrolyzing these substances. Information from these studies should then provide the tools to answer more specific questions as to the function of the individual enzymes at specific sites.

## Degradation of Thyrotropin-Releasing Hormone (TRH)

The rapid degradation of TRH by blood and serum enzymes was already observed before the isolation of the tripeptideamide. Since it was also known that TRH is not degraded by general proteolytic enzymes, such as aminopeptidases, carboxypeptidase B, trypsin, chymotrypsin, pepsin, subtilisin etc., it has been postulated that TRH is degraded by TRH specific enzymes. Preliminary studies on the fragmentation of TRH by tissue enzymes supported this hypothesis and, furthermore, suggested a regulatory role for TRH-metabolizing enzymes (Prasad and Peterkofsky 1976). However, subsequent studies on the catabolism of TRH by tissue enzymes (Bauer and Kleinkauf 1980), did not support this concept and indicated that TRH is degraded by enzymes with broader substrate specificity.

Extensive enzyme characterization studies then clearly demonstrated that the deamidation of TRH is not catalyzed by a specific "TRH-Deamidase" but by a post-proline cleaving enzyme (Knisatschek and Bauer 1979, Hersh and McKelvey 1979) which had been originally characterized by Koida and Walter (1976). As shown by several authors, this enzyme hydrolyzes a variety of neuropeptides at internal proline-X bonds (Orlowksi et al. 1979, Taylor and Dixon 1980, Carvalho and Camargo 1981).

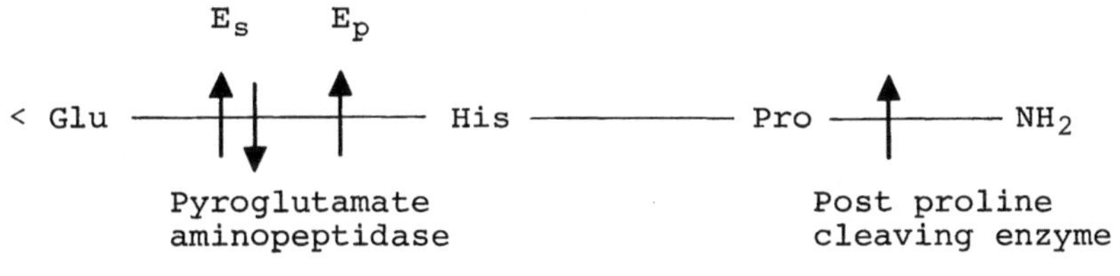


The hydrolysis of TRH at the pyroGlu-His bond is catalyzed by an enzyme which had been characterized before as a pyroglutamate aminopeptidase (Armentrout and Doolittle 1968, Szewczuk and Kwiatkowsky 1970, Mudge and Fellows 1974). This enzyme hydrolyzes a variety of pyroglutamyl-containing peptides such as LH-RH, neurotensin, gastrin as well as synthetic substrates specifically at the pyroGlu-X bond.

Subcellular fractionation studies demonstrated that both enzymes are almost exclusively present in the cytosolic fraction (Horsthemke et al. 1984a,b). In addition, these studies also revealed the existence of another, plasma-membrane bound enzyme (Ep) which hydrolyzes TRH at the pyroGlu-His bond. This enzyme seems to be identical to the peptidase described by Greaney et al. (1980), but it differs from the cytosolic pyroglutamate aminopeptidase in various physical and chemical properties. For example, this enzyme of high molecular weight (280,000 compared to 28,000 for the pyroglutamate aminopeptidase) is inhibited by dithiothreitol and EDTA (Horsthemke et al. 1984a,b, O'Connor and O'Cuinn 1984), while the pyroglutamate aminopeptidase is activated by these agents.

Previously, a TRH-degrading serum enzyme ($E_s$) was characterized which exhibits almost identical characteristics (Taylor and Dixon 1978, Bauer and Nowak 1979). Moreover and of most importance, the TRH-degrading serum enzyme (Bauer and Nowak 1980) and the plasma membrane-bound enzyme (O'Connor and O'Cuinn 1984) exhibit a high degree of substrate specificity. Unlike the pyroglutamate aminopeptidase, these enzymes do not degrade LH-RH, neurotensin, and other pyroglutamyl-containing peptides. It seems possible that these activities represent membrane-bound and secreted forms of the same gene product. Further studies are clearly needed to evaluate the relationship between these enzymes.

The observation that in the rat the activity of the TRH-degrading serum enzyme seems to be controlled by thyroid hormones (Bauer 1976, White et al. 1976) and drastically alters with developmental changes (Oliver et al. 1977, Neary et al. 1976), furthermore suggests that these enzymes might serve very specific functions. Hypothetically, the serum enzyme may represent a functional control element since degradation during transport by the hypophyseal portal blood might influence the amount of TRH, which becomes available at the trophic cells of the pituitary.

The membrane-bound TRH-degrading enzyme might potentially fulfill a specific function for the degradation of TRH at the target sites. In this context it is noteworthy that TRH has also been found in extra-hypothalamic brain areas. Since high affinity receptors for TRH have been demonstrated in these structures, the existence of an effective inactivation mechanism would be prerequisitory for the suggested function of TRH as a neuromodulator or neurotransmitter at these sites. Recent studies with murine cells in primary culture support this hypothesis. After addition of radiolabeled TRH to the culture media, we observed that TRH is rapidly degraded by neuronal cells, but only very slowly by glial cells in primary culture (Schulz and Bauer, unpublished). First of all, these results demonstrate that this TRH-degrading enzyme is localized at the extracellular side of the plasma membrane. Furthermore, the heterogeneous distribution of this enzyme between neuronal and glial cells indicates that this TRH-degrading enzyme is not a general, membrane-constituent protein and, therefore, suggests that it may serve more specialized functions. Of course, further studies are required to evaluate the biological function of the TRH-degrading serum and plasma-membrane bound enzymes.

## Degradation of Luteinizing Hormone-Releasing Hormone (LH-RH)

The rapid degradation of LH-RH by brain and pituitary homogenates has been attributed to the action of a LH-RH specific peptidase. Subsequently, it has been suggested on the basis of very preliminary studies that the activity of this enzyme is regulated by steroid hormones through feedback-regulatory mechanisms and therefore it has been postulated that this enzyme might act as a functional control element (Griffiths et al. 1975, Fridkin et al. 1977, Advis et al. 1982). The characterization of the TRH-degrading enzymes already demonstrated, however, that LH-RH is not only degraded by one enzyme but is also subject to degradation by the pyroglutamate aminopeptidase ($E_1$) and the post-proline cleaving enzyme ($E_2$).

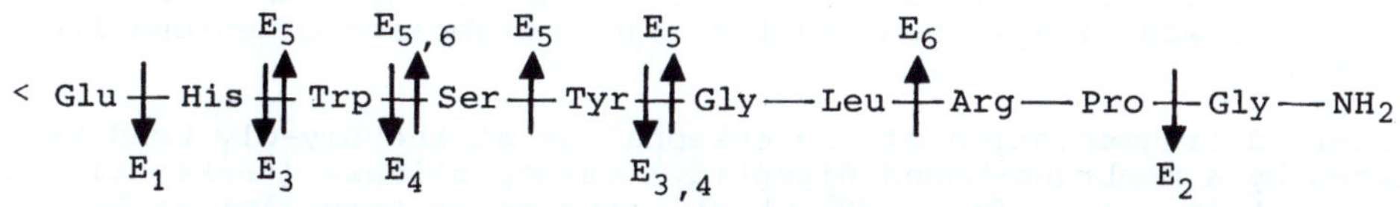


Meanwhile, other enzymes capable of degrading LH-RH have also been identified, namely a nonchymotrypsin-like endopeptidase ($E_3$) (Horsthemke and Bauer 1980, 1982), a cation-sensitive neutral endopeptidase ($E_4$) (Wilk and Orlowski 1980, Horsthemke and Bauer 1981) and a particulate LH-RH degrading enzyme ($E_5$) (Leblanc et al. 1980) which cleave the decapeptideamide at multiple sites. Very recently, it has been reported that LH-RH is also degraded by angiotensin converting enzyme ($E_6$) (Skidgel and Erdös 1985) by hydrolyzing primarily the Trp-Ser and the Leu-Arg peptide bonds. A specific LH-RH degrading enzyme, however, has not yet been identified.

Subcellular fractionation studies with rat adenohypophyseal homogenates demonstrated that the particulate LH-RH degrading enzyme is located in the mitochondrial matrix space (Leblanc et al. 1984), while the other enzymes ($E_{1-4}$) were found in the cytosolic supernatant. From these studies no evidence could be obtained for the existence of a plasmamembrane-bound LH-RH-degrading adenohypophyseal enzyme. Since enzyme activities might have been lost during the subcellular fractionation procedures, we also investigated the degradation of LH-RH by dispersed pituitary cells in culture (in collaboration with Dr. Carl Denef, Leuven, Belgium). In aggrement with the report by Nikolics et al. (1983), and confirmatory to our subcellular fractionation studies, we did not find any significant degradation of LH-RH by intact pituitary cells. These results suggest that at the pituitary level other inactivation mechanisms might exist, for example, the receptor-mediated endocytosis which could be adequate for clearing LH-RH target sites, since LH-RH is known to be released in a pulsatile fashion with very low frequency in the range of 1 - 2 h. Whether this is also true for the inactivation of centrally acting LH-RH remains to be investigated.

## Degradation of Enkephalin

Due to their unprotected amino- and carboxy terminals, enkephalins are subject to extremely rapid degradation by various exopeptidases. This feature accounts for the short half-life of these pentapeptides.

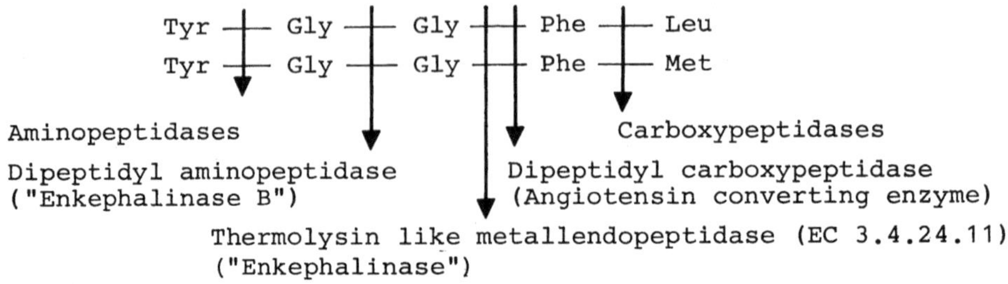


The removal of the aminoterminal tyrosine and thereby the inactivation of the pentapeptides is catalyzed by a variety of as yet not fully characterized aminopeptidase (Hambrook et al. 1975, Hersh and McKelvey 1981, Wagner et al. 1981, Hui et al. 1983). In all systems studied so far, tyrosine was always found as the major metabolite of enkephalin degradation.

The observed fragmentation of the enkephalins at the Gly-Gly bond is catalyzed by a membrane-bound dipeptidyl aminopeptidase ("enkephalinase B") (Gorenstein and Snyder 1979) which seems to be very similar to the cytosolic dipeptidyl aminopeptidase III (Ellis and Nuenke 1967, Lee and Snyder 1982). After purification to homogeneity, we observed almost identical substrate specificity for the two enzymes, the same molecular weight of 83,000 and very similar enzymechemical characteristics (Bergmann and Bauer, unpublished).

The cleavage of the pentapeptides at the Gly-Phe bonds is catalyzed by angiotensin-converting enzyme (Erdös et al. 1978) and by an enzyme which had been termed enkephalinase, since initial studies had indicated that this enzyme exhibits strict substrate specificity and a very high affinity ($K_m$ values in the nM range) towards enkephalins (Malfroy et al. 1978). Furthermore, it had been reported that this enzyme activity increases in the brain of morphine-treated mice, and it has been suggested that this increase results from a feedback-mechanism operating at the target cells. The observation that thiorphan, a synthetic inhibitor of "enkephalinase", showed antinociceptive activity which could be reversed by naloxone, an opiate antagonist, also supported the hypothesis that this enzyme is specifically involved in the regulation of enkephalinergic transmission. Extensive studies by several laboratories demonstrated, however, that "enkephalinase" is apparently identical to a thermolysin-like neutral metallendopeptidase (EC 3.4.24.11) (Almenoff et al. 1981, Fulcher et al. 1982) which had been originally identified by Kerr and Kenney already in 1974.

Subcellular fractionation studies demonstrated that enkephalin-degrading aminopeptidases, the dipeptidyl aminopeptidase, the thermolysin-like metalloendopeptidase and the angiotensin-converting enzyme are all associated with the plasma membrane. In a broad sense, all these enzymes could be regarded as potential "enkephalinases". However, by screening neuronal and glial cells in primary culture for enkephalin-degrading activities, we found only trace amounts of angiotensin-converting enzymes with these cultures (Horsthemke et al. 1983). In agreement with this result is the observation that angiotensin-converting enzyme is mainly associated with endothelial cells (Johnson and Erdös 1978). The rapid degradation of Leu-enkephalin by glial and neuronal cells in primary culture is mainly due to bestatin-sensitive aminopeptidases. The dipeptidyl aminopeptidasic activity is much higher in the neuronal than in the glial cultures, whereas the opposite is true for the thermolysin-like metalloendopeptidase. Similar results have been reported by Lentzen and Palenker (1983).

Although the structural and functional integrity of cell interaction is not conserved in primary cultures, the present result suggest that in the case of the enkephalins the bestatin-sensitive aminopeptidases on nerve cells could potentially be involved in the removal of enkephalins from the synaptic cleft, whereas the "enkephalinase" on glial cells or angiotensin converting enzyme on endothelial cells may hydrolyze peptides which already left the synaptic cleft by diffusion and therefore may serve general scavenger functions.

## Degradation of Substance P

Substance P is widely distributed in the central and peripheral nervous system and seems to be involved in neurotransmission and other forms of cellular communication. Little is known about the inactivation of substance P, although a number of enzymes capable of degrading this peptide have been reported to exist in brain.

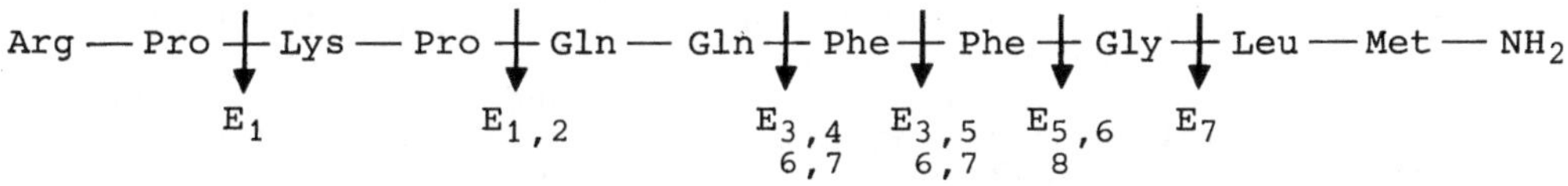


Sequential removal of Arg-Pro and Lys-Pro residues by the dipeptidyl aminopeptidase IV ($E_1$; post-proline dipeptidyl aminopeptidase) leads to the C-terminal heptapeptideamide which retains full biological activity. The same fragment is formed by the cytosolic post-proline cleaving endopeptidase ($E_2$) and, therefore, it seems unlikely that these substance P-degrading enzymes contribute to the inactivation of the undecapeptideamide.

Akopyan et al. (1979) described a cytosolic neutral endopeptidase ($E_3$) which hydrolyzes substance P, probably between Gln-Phe and Phe-Phe bonds. Purified lysosomal cathepsin D ($E_4$) (Benuck et al. 1977) cleaves the undecapeptideamide between Phe-Phe and a cytosolic enzyme called tonin ($E_5$) (Chrêtien et al. 1980) was reported to hydrolyze the Phe-Phe and the Phe-Gly peptide bonds. The intracellular localization of these enzymes makes it unlikely that they participate in the inactivation of extracellular substance P.

The same seems true for a particle bound substance P degrading enzyme recently isolated from human brain ($E_6$) (Lee et al. 1981). This neutral endopeptidase of 40,000 - 50,000 molecular weight appears to exhibit a high specificity towards substance P, while cleaving the undecapeptideamide at multiple sites (Gln-Phe, Phe-Phe, and Phe-Gly). Subcellular fractionation studies indicated, however, that this enzyme is localized in the mitochondria (Horsthemke et al. 1984a,b).

Degradation of substance P is effectively catalyzed by the thermolysin-like metalloendopeptidase ("enkephalinase") ($E_7$) (Matsas et al. 1983) which hydrolyzes the bonds Gln-Phe, Phe-Phe and Gly-Leu. As mentioned before, however, this enzyme appears to be preferentially associated with glial cells in primary cultures and therefore it seems less likely that the thermolysin-like metalloendopeptidase may play an important role in the inactivation of substance P in the synaptic cleft.

Recent studies with purified angiotensin-converting enzyme demonstrated that substance P is also degraded by angiotensin-converting enzyme ($E_8$)

by hydrolyzing primarily the Phe-Gly bond (Yokosawa et al. 1983, Skidgel et al. 1984). Based on extensive studies with a variety of peptide substrates, this enzyme was characterized as a dipeptidyl carboxypeptidase, but the present results provide strong evidence that angiotensin-converting enzyme may also exhibit endopeptidasic activities. Due to its preferential association with endothelial cells, however, angiotensin-converting enzyme may fulfill general scavenger functions, but does not seem to be a potential candidate for an enzyme serving specialized target site clearance functions.

For the reasons just mentioned, it seems unlikely that the enzymes characterized so far participate in inactivating extracellular substance P in the synaptic cleft. This notion is supported by our studies on the degradation of substance P by glial and neuronal cells in primary culture (Horsthemke et al. 1984b). With these cells we observed rapid degradation of substance P, which was reduced only by 30% when the culture medium was supplemented with a combination of inhibitors for the aforementioned enzymes. In contrast, the degradation of substance P was inhibited most effectively by the addition of bacitracin as the only enzyme inhibitor. Since the specific activity of this bacitracin-sensitive substance P-degrading enzyme is nearly fivefold higher in the neuronal than in glial cell cultures, these results suggest that this enzyme is probably the best candidate as a peptidase likely to be involved in the synaptic inactivation of substance P. Since the cell cultures do not contain functional synapses, immunohistochemical and cytochemical studies are needed to evaluate this hypothesis and, therefore, it will be very interesting to isolate and to characterize this bacitracin-sensitive substance P-degrading enzyme.

## Conclusions and Reflections

The biochemical studies did not reveal any evidence for the presence of neuropeptide-specific enzymes in the brain or the pituitary. The primary specificity of the enzymes identified in these tissues is rather directed towards certain structural elements of the peptide (e.g., the aminoterminus or an amino acid side chain etc.). This does not exclude the possibility, however, that the rate of hydrolysis of a given peptide bond is strongly influenced by neighboring groups, the size and the conformation of the substrates. The $K_m$-values are in the micromolar range, as expected for general peptidases.

Potentially, such enzymes are capable of degrading an unlimited number of peptides provided they contain the appropriate structural elements. It is clear, however, that these enzymes, which lack the specificity and selectively required for a regulation process, cannot fulfill a regulatory function themselves. Therefore, the concept that neuropeptide-degrading enzymes directly serve a dynamic control function cannot be correct. The fluctuations in enzyme activities which were observed after physiological manipulations of experimental animals are unlikely to reflect specific changes within feedback-regulatory mechanisms, but are rather due to changes within homeostatic adaption processes (Loudes et al. 1978, Bauer et al. 1981a,b).

The biochemical studies cannot provide any direct information as to the biological function of the individual peptidases. Since a given peptidase potentially can act on several peptides and, vice versa, a given peptide could be degraded by several peptidases, such a degradation system is obviously most suitable for serving general scavenger functions. The more specific target site clearance function of such enzymes, and eventually their function as local converting enzymes, is primarily governed by the strategic localization of the enzyme in relation to a given peptidergic pathway. Since we cannot expect that a given enzyme is exclusively concerned with the degradation of only one peptide, we cannot expect that the regional distribution of the enzyme corresponds fully with a specific peptidergic pathway and, therefore, it will be most important to determine at the electron microscope level the exact localization of the individual enzyme *in relation* to a given peptidergic terminal. Only at this level might we find some information as to the biological function of a given peptidase for the inactivation of a specific peptide at a specific site.

The tools for these investigations have to be provided by biochemical studies. For example, the purification of the peptidases to homogeneity should enable us to raise enzyme-specific antibodies which can be used for immunohistochemical studies. For this, the application of the immunogold technique might be most appropriate to determine the localization of these enzymes at the electron microscope level. In selected cases, enzyme-specific substrates could also be used for such studies. This approach would primarily depend on the properties of the enzymes which have to be evaluated before.

Extensive characterization studies of neuropeptide-degrading enzymes will hopefully also provide the basis for the development of other areas of neuropeptide research. For example, these studies should offer valuable clues for the design and synthesis of enzyme-resistant analogs suitable for physiological and pharmacological investigations. The elucidation of the substrate specificity and catalytic mechanisms of neuropeptide-degrading enzymes should finally also provide the basis for the synthesis of enzyme-specific inhibitors, which could be extremely useful for pharmacological studies. This approach has been successfully used already for the synthesis of highly potent and selective inhibitors of angiotensin-converting enzyme, for example. Some of these inhibitors, such as captopril, are already clinically used as antihypertensive drugs. The effect of such inhibitors in whole animals are in agreement with the concept that in the periphery this enzyme is involved in controlling the blood pressure.

However, the effects of centrally acting inhibitors of neuropeptide-degrading enzymes might be more difficult to evaluate. Although the target site clearance function of neuropeptide-degrading enzymes is primarily determined by the strategic localization of the enzyme in relation to a given peptidergic pathway, we may expect that at different peptidergic terminals the same enzyme could be involved in the inactivation of different peptides and, therefore, severe side-effects might be anticipated even when specific enzyme-inhibitors are used. Therefore, it remains to be investigated whether therapeutical useful inhibitors of neuropeptide-degrading enzymes could be developed. Nevertheless, for basic research, these inhibitors could be very interesting and important pharmacological tools. It is hoped that studies along such lines will not only help to attain an understanding on the function of neuropeptide-degrading enzymes, but, hopefully, will also contribute to further our knowledge on the biological function of neuropeptides in general.

*Acknowledgment.* Our work reported here was supported by the Deutsche Forschungsgemeinschaft.

## References

Advis JP, Krause JE, McKelvey JF (1982) Evidence that endopeptidase-catalyzed luteinizing hormone releasing hormone cleavage contributes to the regulation of median eminence LHRH levels during positive steroid feedback. Endocrinology 112: 1147-1149

Akopyan TN, Arutunyan AA, Organisyan AI, Lajtha A, Galoyan AA (1979) Breakdwon of hypothalamic peptides by hypothalamic neutral endopeptidase. J Neurochem 32:629-631

Almenoff J, Wilk S, Orlowski M (1981) Membrane bound pituitary endopeptidase: apparent identity to enkephalinase. Biochem Biophys Res Commun 102:206-214

Armentrout RM, Doolittle RF (1968) Pyrrolidonecarboxylyl peptidase: stabilization and purification. Arch Biochem Biophys 132:80-90

Bauer K (1976) Regulation of degradation of thyrotropin releasing hormone by thyroid hormones. Nature (London) 259:591-593

Bauer K, Kleinkauf H (1980) Catabolism of thyroliberin by rat adenohyophyseal tissue extract. Eur J Biochem 106:107-117

Bauer K, Nowak P (1979) Characterization of a thyroliberin-degrading serum enzyme catalyzing the hydrolysis of thyroliberin at the pyroglutamyl-histidine bond. Eur J Biochem 99:239-246

Bauer K, Nowak P, Kleinkauf H (1981a) Specificity of a serum peptidase hydrolyzing thyroliberin at the pyroglutamyl-histidine bond. Eur J Biochem 118:173-176

Bauer K, Beier S, Horsthemke B, Knisatschek H, Sievers J (1981b) Estrogen effects on LH-RH degrading brain and pituitary enzymes. Exp Brain Res Suppl 3:93-107

Benuck M, Grynbaum A, Marks N (1977) Breakdown of somatostatin and substance P by cathepsin D purified from calf brain by affinity chromatography. Brain Res 143: 181-185

Carvalho KM, Camargo ACM (1981) Purification of rabbit brain endo-oligopeptidases and preparation of anti-enzyme antibodies. Biochemistry 20:7082-7088

Chrêtien M, Lee CM, Sandberg BEB, Iversen LL, Boucher R, Seidah NG, Genest J (1980) Substrate specificity of the enzyme tonin: Cleavage of substance P. FEBS Lett 113:173-176

Ellis S, Nuenke JM (1967) Dipeptidyl arylamidase III of the pituitary. J Biol Chem 242:4623-4629

Erdös EG, Johnson AR, Boyden NT (1978) Hydrolysis of enkephalin by cultured human endothelial cells and by purified peptidyl dipeptidase. Biochem Pharmacol 27: 843-848

Fridkin M, Hazum E, Baram T, Lindner HR, Koch Y (1977) Hypothalamic and pituitary LRF-degrading enzymes: characterization, purification and physiological role. In: Goodman M, Meienhofer J (eds) Proc 5th Am Peptide Symp. Halsted Press, New York, pp 193-196

Fulcher IS, Matsas R, Turner AJ, Kenney AJ (1982) Kidney neutral endopeptidase and the hydrolysis of enkephalin by synaptic membranes show similar sensitivity to inhibitors. Biochem J 203:519-522

Gorenstein C, Snyder SH (1979) Two distinct enkephalinases: Solubilization, partial purification and separation from angiotensin converting enzyme. Life Sci 25: 2065-2070

Greaney A, Phelan J, O'Cuinn G (1980) Localization of thyroliberin pyroglutamyl peptidase on synaptosomal-membrane preparations of guinea-pig brain tissue. Biochem Soc Trans 8:423

Griffiths EC, Hooper KC, Jeffcoate SL, Holland DT (1975) The effects of gonadectomy and gonadal steroids on the activity of hypothalamic peptidases inactivating luteinizing hormone-releasing hormone (LH-RH). Brain Res 88:384-388

Hambrook JM, Morgan BA, Rance MJ, Smith CF (1976) Mode of deactivation of the enkephalins by rat and human plasma and rat brain homogenates. Nature (London) 262:782-783

Hersh LB, McKelvey JF (1979) Enzymes involved in the degradation of thyrotropin releasing hormone (TRH) and luteinizing hormone releasing hormone (LH-RH) in bovine brain. Brain Res 168:553-564

Hersh LB, McKelvey JF (1981) An aminopeptidase from rat brain which catalyzes the hydrolysis of enkephalin. J Neurochem 13:171-178

Horsthemke B, Bauer K (1980) Characterization of a non-chymotrypsin-like endopeptidase from anterior pituitary that degrades luteinizing hormone-releasing hormone at the tyrosyl-glycine and histidyl-tryptophan bonds. Biochemistry 19:2867-2873

Horsthemke B, Bauer K (1981) Chymotryptic-like hydrolysis of luliberin (LH-RF) by an adenohypophyseal enzyme of high molecular weight. Biochem Biophys Res Commun 103:1322-1328

Horsthemke B, Bauer K (1982) Substrate specificity of an adenohypophyseal endopeptidase capable of hydrolyzing luteinizing hormone-releasing hormone: Preferential cleavage of peptide bonds involving the carboxyl terminus of hydrophobic and basic amino acids. Biochemistry 21:1033-1036

Horsthemke B, Hamprecht B, Bauer K (1983) Heterogenous distribution of enkephalin-degrading peptidases between neuronal and glial cells. Biochem Biophys Res Commun 115:423-429

Horsthemke B, Leblanc P, Kordon C, Wattiaux-de Coninck S, Wattiaux R, Bauer K (1984a) Subcellular distribution of particle-bound neutral peptidases capable of hydrolyzing gonadoliberin, thyroliberin, enkephalin and substance P. Eur J Biochem 139:315-320

Horsthemke B, Schulz M, Bauer K (1984b) Degradation of substance P by neurones and glial cells. Biochem Biophys Res Commun 125:728-733

Hui KS, Wang YJ, Lajtha A (1983) Purification of an enkephalin aminopeptidase from rat brain membranes. Biochemistry 22:1062-1067

Johnson AR, Erdös EG (1978) Metabolism of vasoactive peptides by human endothelial cells in culture. J Clin Invest 59:684-689

Kerr MA, Kenney AJ (1974) The purification and specificity of a neutral endopeptidase from rabbit kidney brush border. Biochem J 137:477-488

Knisatschek H, Bauer K (1979) Characterization of 'thyroliberin-deamidating enzyme' as a post-proline cleaving enzyme. J Biol Chem 254:10936-10943

Koida M, Walter R (1976) Post-proline cleaving enzyme. J Biol Chem 251:7593-7599

Leblanc P, Patton E, L'Heritier, Kordon C (1980) Some properties of peptidasic activity bound to the anterior pituitary membranes. Biochem Biophys Res Commun 96:1457-1465

Leblanc P, L'Heritier A, Kordon C, Horsthemke B, Bauer K, Wattiaux-de Coninck S, Dubois F, Wattiaux R (1984) Characterization of a neutral endopeptidase localized in the mitochondrial matrix of rat anterior pituitary tissue with GnRH as substrate. Neuroendocrinology 38:476-483

Lee CM, Snyder SH (1982) Dipeptidylaminopeptidase III of rat brain. J Biol Chem 257:12043-12050

Lee CM, Sandberg BEB, Hanley MR, Iversen LL (1981) Purification and characterization of a membrane-bound substance P-degrading enzyme from human brain. Eur J Biochem 114:315-327

Lentzen H, Palenker I (1983) Localization of the thiorphan-sensitive endopeptidase, termed enkephalinase A, on glial cells. FEBS Lett 153:93-97

Loudes C, Josepha-Bravo P, Leblanc P, Kordon C (1978) Specific activity of LHRH and TRH degrading enzymes in various tissues of normal and castrated male rats. Biochem Biophys Res Commun 83:921-926

Malfroy B, Swerts JP, Guyon A, Roqes BP, Schwartz JC (1978) High affinity enkephalin-degrading peptidase in brain is increased after morphine. Nature (London) 276: 523-526

Matsas R, Fulcher IS, Kenney AJ, Turner AJ (1983) Substance P and Leu-enkephalin are hydrolyzed by an enzyme in pig caudate synaptic membranes that is identical with the endopeptidase of kidney microvilli. Proc Natl Acad Sci USA 80:3111-3115

Mudge AW, Fellows RE (1973) Bovine pituitary pyrrolidonecarboxyl peptidase. Endocrinology 93:1428-1434

Mumford RA, Pierczchala PA, Strauss AW, Zimmermann M (1981) Purification of a membrane-bound metalloendopeptidase from porcine kidney that degrades peptide hormones. Proc Natl Acad Sci USA 78:6623-6627

Neary JT, Kiefer JD, Federico P, Mover H, Maloof F (1976) Thyrotropin releasing hormone: Development of inactivation system during maturation of the rat. Science 193:403-405

Nikolics K, Szoke B, Keri G, Teplan I (1983) Gonadotropin releasing hormone (GnRH) is not degraded by intact pituitary tissue in vitro. Biochem Biophys Res Commun 114:1028-1035

O'Connor B, O'Cuinn G (1984) Localisation of a narrow specificity thyroliberin hydrolyzing pyroglutamate aminopeptidase in synaptosomal membranes of guinea pig brain. Eur J Biochem 144:271-278

Oliver C, Parker CR, Porter JC (1977) Developmental changes in the degradation of thyrotropin releasing hormone by the serum and brain tissues of the male rat. J Endocrinol 74:339-340

Orlowski M, Wilk E, Pierce S, Wilk S (1979) Purification and properties of a prolyl endopeptidase from rabbit brain. J Neurochem 33:461-469

Prasad C, Peterkofsky A (1976) Demonstration of pyroglutamylpeptidase and amidase activities toward thyrotropin-releasing hormone in hamster hypothalamus extract. J Biol Chem 251:3229-3234

Skidgel RA, Erdös EG (1985) Novel activity of human angiotensin I converting enzyme: Release of the $NH_2$- and COOH-terminal tripeptides from the luteinizing-hormone releasing hormone. Proc Natl Acad Sci USA 82:1025-1029

Skidgel RA, Engelbrecht S, Johnson AR, Erdös EG (1984) Hydrolysis of substance P and neurotensin by converting enzyme and neutral endopeptidase. Peptides 5: 769-776

Szewczuk A, Kwiatkowsky J (1970) Pyrrolidonyl peptidase in animal, plant and human tissues. Eur J Biochem 15:92-96

Taylor WL, Dixon JE (1978) Characterization of a pyroglutamate aminopeptidase from rat serum that degrades thyrotropin-releasing hormone. J Biol Chem 253:6934-6940

Taylor WL, Dixon JE (1980) Catabolism of neuropeptides by a brain proline endopeptidase. Biochem Biophys Res Commun 94:9-15

Wagner GW, Tavianini MA, Herrmann KM, Dixon JE (1981) Purification and characterization of an enkephalin degrading aminopeptidase from rat brain. Biochemistry 20: 3884-3890

White N, Jeffcoate SL, Griffiths EC, Hooper KC (1976) Effect of thyroid status on the thyrotropin-releasing hormone-degrading activity of rat serum. J Endocrinol 71:13-19

Wilk S, Orlowski M (1980) Cation-sensitive neutral endopeptidase: isolation and specificity of the bovine pituitary enzyme. J Neurochem 35:1172-1182

Yokosawa H, Endo S, Ogura Y, Ishii S (1983) A new feature of angiotensin-converting enzyme in the brain: Hydrolysis of substance P. Biochem Biophys Res Commun 116: 735-742

# Structure and Function of Cholinergic Synaptic Vesicles

H. Stadler, M.-L. Kiene, P. Harlos, and U. Welscher[1]

## Introduction

The function of a nerve terminal is to release neurotransmitter to transfer a signal to a receptor cell. The nerve terminal is a specialized region of the neuron, it is characterized by the presence of numerous membrane-bound organelles, the synaptic vesicles. The vesicles store and release the neurotransmitter and thus play a central role in synaptic transmission. The nerve terminals are separated from the cell body by their long and thin axon and the neuron therefore represents an extremely polarized cell. In the axon and in the nerve terminal there is no protein synthesis, and in order to maintain synaptic function the nerve terminal has to be supplied continuously with synaptic vesicle membranes and other material from the cell body. In addition to the nerve terminal, the cell body has to supply the dendrites. As a consequence the problem of membrane sorting is especially intriguing and challenging in the neuron (Fig. 1).

For the biosynthesis and membrane sorting of synaptic vesicles we may distinguish four steps:

1. Synaptic vesicle proteins must be segregated away from dendritic and plasma membrane proteins and be correctly assembled.
2. The synaptic vesicle membrane must be targeted to be transported by the axon to the nerve terminal.
3. In the nerve terminal the synaptic vesicle must recognize a specialized domain of the presynaptic plasma membrane where it fuses.
4. Since transmitter is released in juxta-position to the receptor, a transsynaptic mechanism should exist that keeps pre- and postsynaptic structures together.

In addition the synaptic vesicle must at some stage be filled with transmitter.

The results presented here are related to topics (3) and (4) including axonal transport of synaptic vesicles, identification of vesicle-specific proteins, uptake, storage and exocytosis of transmitter.

Our model system is the cholinergic electromotoneurone of the electric ray *Torpedo marmorata* (Fig. 2). The electric organs of *Torpedo* provide an excellent source of single transmitter type synaptic vesicles. In addition, the isolation of electromotoneuron axons and cell bodies can be carried out easily.

---

[1]Max-Planck-Institut für Biophysikalische Chemie, Abteilung Neurochemie, D-3400 Göttingen, FRG

36. Colloquium - Mosbach 1985
Neurobiochemistry

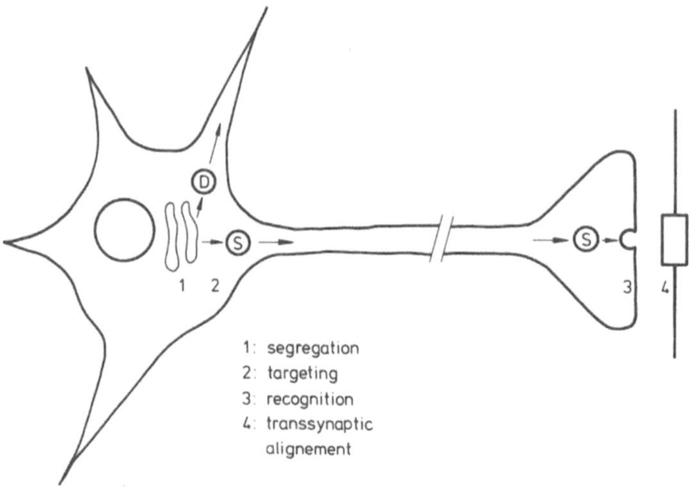


Fig. 1. Schematic representation of vesicle membrane sorting in the neuron (explanation see text). *D* dendritic transport; *S* synaptic vesicle

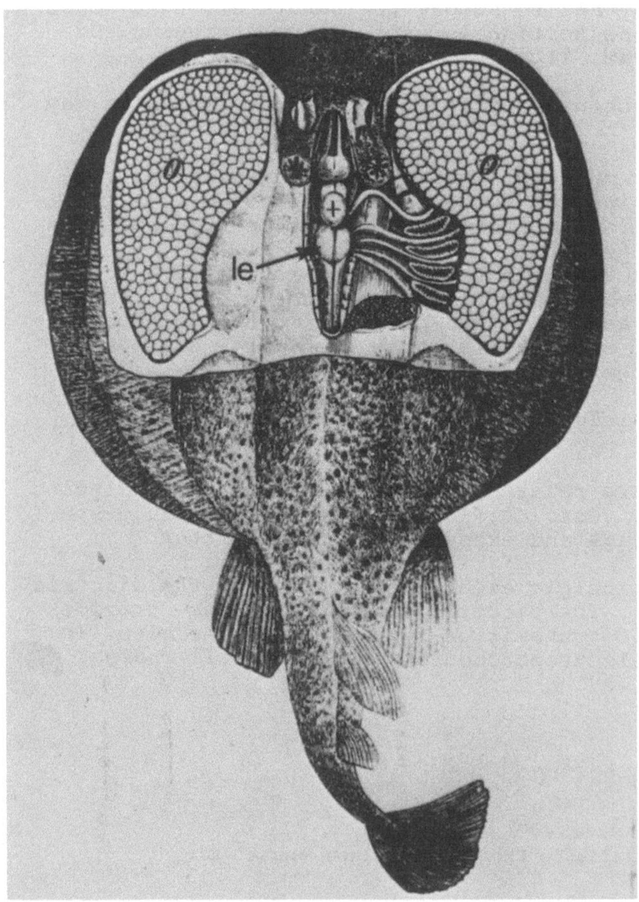


Fig. 2. The electric ray *Torpedo marmorata*. *O* electric organ; *le* lobus electricus. (Drawing by Fritsch 1890)

## Methods

Synaptic vesicles and synaptosomal plasma membranes were isolated from electric organs of *Torpedo marmorata* as described previously (Tashiro and Stadler 1978, Stadler and Tashiro 1979). The identification of a vesicular nucleotide carrier is outlined in Stadler and Fenwick (1983), of a vesicular proteoglycan in Stadler and Dowe (1982) and of a vesicular proton pump in Harlos et al. (1984). $^{1}H$-NMR analysis of synaptic vesicles is described in Stadler and Füldner (1980), determination of intravesicular pH in Stadler and Füldner (1981), and a detailed $^{31}P$-NMR analysis of synaptic vesicles and electric organ tissue is reported in Füldner and Stadler (1982). The production of a vesicle-specific antiserum is described in Walker et al. (1982), immunofluorescence procedures in Jones et al. (1982). Isolation of synaptic vesicles from guinea pig brain was as described (Stadler and Tsukita 1984).

An antiserum against a vesicular proteoglycan has been prepared by injection four times 100 - 200 µg purified proteoglycan at 14-day intervals into a rabbit. A quantitative proteoglycan determination has been carried out using this antiserum in a dot blot assay. Material from subcellular fractionation is concentrated by TCA precipitation, dissolved in SDS-containing buffer and spotted on nitrocellulose paper. After incubation with the antiserum and various wash steps, the nitrocellulose paper is incubated with $^{125}I$-labeled anti-rabbit IgG antiserum (Amersham, Braunschweig) and after further washing the spots are counted for radioactivity (Kiene and Stadler, in preparation).

For axonal transport experiments the lobi electrici of *Torpedo marmorata* were injected with 1-2 mCi $^{35}S$-sulfate and axons and electric organs were removed for further analysis after various time intervals. Isolation of radiolabeled vesicles from this material was as described (Tashiro and Stadler 1978).

## Results and Discussion

### Isolation of Synaptic Vesicles

Synaptic vesicles appear in the electric organ of *Torpedo* as membrane-bound spherical organelles of about 90 nm diameter.

Whittaker and coworkers were the first to isolate synaptic vesicles from the organ (Whittaker et al. 1972), and showed that beside acetylcholine ATP is present in these organelles (Dowdall et al. 1974). Later we succeeded in obtaining a highly purified fraction (Tashiro and Stadler 1978) and it became clear that the vesicles store acetylcholine in concentrations as high as 1 M and ATP as high as 0.1 - 0.2 M (Ohsawa et al. 1979). This raises the question of how the vesicles accumulate these enormous amounts, and therefore the underlying storage mechanisms and uptake systems were investigated.

### NMR Analysis of the Storage Mechanism

The high concentration of acetylcholine and ATP suggested that these molecules are not in solution but probably immobilized or interacting with charged macromolecules to keep isoosmolarity of the granule to *Torpedo* tissue which is around 900 $mOsmol \cdot l^{-2}$.

High resolution $^{1}H$-NMR analysis of isolated vesicles does not support this hypothesis. The spectral characteristics of the acetylcholine methyl groups clearly indicate that the molecule is fully in solution (Stadler and Füldner 1980).

$^{31}P$-NMR analysis showed that the vesicular ATP has unusual chemical shifts and linebroadening of the phosphate resonances.

These spectral characteristics were explained as exchange broadening of ATP with $Mg^{2+}$ ions and a protonated form of ATP corresponding to a pH of around 5.5 (Füldner and Stadler 1982). It is possible to identify vesicular ATP in the spectrum of pieces of electric organ with the same spectral characteristics as found in the isolated vesicle. The results therefore apply to the in vivo situation underlining the value of the non-invasive NMR technique.

In summary we conclude that ACh and ATP are stored inside the vesicle essentially in free solution at an acidic pH. No evidence for complex formation or immobilization has been found. Concentrated salt solutions like the interior of vesicles are known to have low activity coefficients for the solutes. Therefore there is no reason to postulate hyperosmolarity for the vesicle interior suggested by simply taking concentrations of solutes.

## Membrane Proteins

How are acetylcholine and ATP accumulated, and what could provide the energy for this uptake? It has been found by other groups and our own that acetylcholine and ATP can be taken up in vitro by carrier-mediated processes (Luqmani 1981, Anderson et al. 1982). In the case of ATP, uptake is inhibited by the plant alkaloid atractyloside (Luqmani 1981), a well-known inhibitor of the mitochondrial ADP/ATP carrier. After labeling vesicle membranes with $^{3}H$-atractyloside, we succeeded in solubilizing an atractyloside/protein complex with Triton X-100 from the membranes.

Isoelectric focusing of the material then showed that the protein in the complex corresponds to a major vesicle component. A comparison of the vesicle protein, which we concluded to represent the nucleotide carrier, with the mitochondrial carrier (see Table 1) showed several

Table 1. Comparison of nucleotide carrier properties

| | Mitochondria | Vesicles |
|---|---|---|
| MW (kd) | 30 | 34 |
| IEP | 8-10 | 7-8 |
| Inhib. atractyl. | + | + |
| Signal Sequence | - | -[a] |
| Type | Antiport | Uniport? |

[a]Detected by in vitro translation of mRNA from electric lobe of *Torpedo marmorata* in reticulocyte lysate system (Stadler, unpublished)

similarities suggesting that the two nucleotide carriers have evolved from a common gene precursor (Stadler and Fenwick 1983).

It has not yet been possible to identify the acetylcholine carrier of the vesicle membrane, due to the lack of a specific inhibitor forming a stable complex with the presumed carrier.

## Proton Pump

Since the molecular basis of the ΔpH over the vesicle membrane was not known, we have investigated whether this is created by a proton pump. We found, confirming earlier results (Breer et al. 1977), that vesicles are associated with a Mg-ATPase. This Mg-ATPase can be split off from the vesicle membrane, like the $F_1$-ATPase of submitochondrial particles, with chloroform or dichloromethane and eluted upon gel filtration with about 250 kd, suggesting that the vesicles contain an $F_1$-like ATPase.

Contrary to the $F_1$-ATPase, it is oligomycin-insensitive and contains only one subunit in the 50 kd range (Harlos et al. 1984).

ATP-dependent proton pumping was demonstrated by the technique of $^{14}C$-methylamine; in the presence of protonophores no uptake was observed, indicating that the acidification is not due to a Donnan effect (Harlos et al. 1984).

We recently succeeded in obtaining a highly purified fraction of synaptic vesicles from guinea pig brain. This synaptic vesicle preparation also shows features characteristic for ATP-dependent proton pumping. Similar to the cholinergic vesicle from *Torpedo* we found an $F_1$-like ATPase and ATP-dependent acidification of the core.

We could further show in this case that the vesicles contain protrusions on their surface known to be characteristic for proton ATPases (Stadler and Tsukita 1984).

We conclude that synaptic vesicles in general are equipped with proton ATPases which create a membrane potential that is essential for uptake and storage of neurotransmitter.

A $Ca^{2+}$ uptake system of cholinergic vesicles has been described by other groups (Michaelson et al. 1980). This is a $Ca^{2+}$ uptake system that probably removes inflowing $Ca^{2+}$ after depolarization of the nerve terminal. The uptake system is stimulated by calmodulin (Rephaeli and Parsons 1982). We have identified a $Ca^{2+}$/calmodulin-binding protein in vesicles. It is in the molecular weight range around 140 kd like $Ca^{2+}$-ATPases of other systems and could therefore represent a candidate for the vesicular $Ca^{2+}$ uptake system (Walker et al. 1984). Calmodulin has two further binding sites in the presence of $Ca^{2+}$ as demonstrated by the calmodulin gel overlay technique, suggesting a multifunctional role for the vesicle.

## A Proteoglycan in the Core of Vesicles

An antigenic component of vesicles is a proteoglycan (Stadler and Dowe 1982). Due to the relatively low amounts of total protein (< 500 μg) that we can obtain from one preparation this was difficult to identify, and we had to develop an in vivo labeling procedure. $^{35}S$-sulfate was injected into the lobus electricus of the fish brain where the cell bodies of the electromotoneurons are. After several hours a wave of

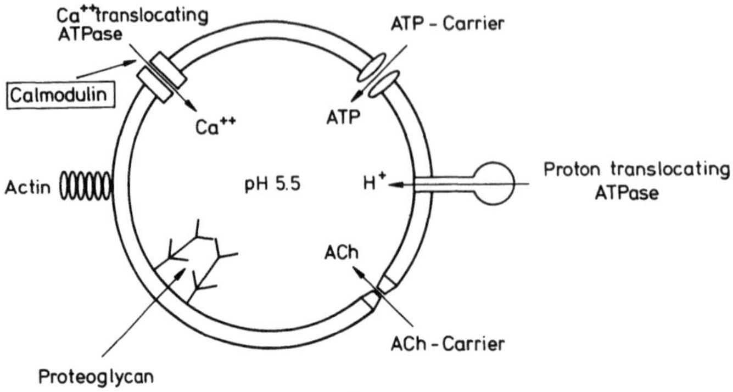


Fig. 3. Diagrammatic representation of our present knowledge of the structure of cholinergic synaptic vesicles from *Torpedo*. The vesicle membrane is associated with carrier systems for acetylcholine, ATP and $Ca^{2+}$. The $Ca^{2+}$-uptake system is stimulated by calmodulin. A proton translocating ATPase provides the energy for uptake of acetylcholine and ATP. The core is filled with a solution of ACh and ATP at an acidic pH. In addition a heparansulfate proteoglycan is present in the core. A further vesicle constituent is actin

radioactivity is observed in the axons migrating with a rate of 100 - 200 mm/day from the cell body to the electric organ.

Autoradiography of SDS gel electrophoresis of axon segments shows that the radioactivity is largely restricted to a broad band in the top of the separating gel or in the stacking gel. After about 24 h the wave has reached the electric organ. When vesicles are extracted from the organ after 48 h, we find that they have become radioactively labeled and that the radioactivity is restricted, as in the axon segments, to a broad band at the top of the gel or in the stacking gel.

Further analysis showed that the sulfate was incorporated into heparansulfate, which itself was bound to a protein backbone forming a proteoglycan. This proteoglycan is a major component of vesicles and it elutes upon gel filtration in SDS with a molecular weight >350 kd (Stadler and Dowe 1982). An antiserum against vesicles can be made vesicle-specific by appropriate adsorption procedures and the residual antibodies are then directed only against the proteoglycan (Walker et al. 1982). This antiserum does not precipitate intact vesicles; only if vesicles are solubilized in Triton X-100 or osmotically shocked can the proteoglycan be precipitated, indicating that the proteoglycan is indeed located inside the vesicle. We can now summarize our present knowledge of the vesicle structure (Fig. 3).

## Vesicle Dynamics and Metabolic Heterogeneity

A monospecific antiserum against purified proteoglycan was obtained recently. This antiserum provides a vesicle marker independent of ACh and ATP. If vesicles are isolated in our standard way by zonal centrifugation on a shallow sucrose gradient it can be seen that vesicles as detected by ACh and ATP measurements contain reproducibly a shoulder to the heavier side of the gradient suggesting that there are two populations of vesicles (Tashiro and Stadler 1978, cf. Fig. 4a).

The $VP_2$ population is increasing if the electric organ is electrically stimulated and in a series of experiments using the perfused electric

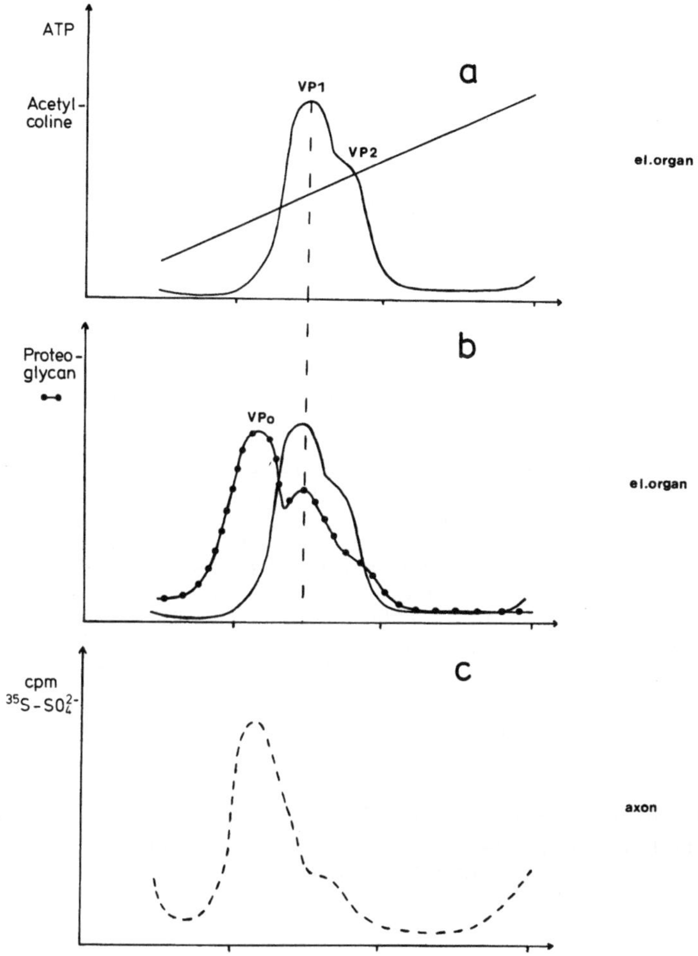


Fig. 4a-c. Metabolic heterogeneity of cholinergic vesicles as a the classical vesicle markers ATP or acetylcholine detect populations $VP_1$ (mature granule) and $VP_2$ (has undergone exo-/endocytotic cycles); b a quantitative immunassay against the vesicular proteoglycan detects population $VP_0$; c vesicles isolated from axons labeled in vivo by $^{35}S$-incorporation comigrate with $VP_0$, indicating that $VP_0$ is the axonally transported population. *Abscissa* eluate of zonal density gradient centrifugation of vesicles isolated from electric organs (a,b) and from axons (c). *Ordinate* classical vesicle marker ATP and acetylcholine (a)

organ it has been shown that this population takes up preferentially newly synthesized transmitter. It must have been in contact with the extracellular fluid at some stage because markers present during electrical stimulation in the perfusate can be found afterwards incorporated into the $VP_2$ fraction (Zimmermann et al. 1981). These results provide evidence that acetylcholine is released from vesicles by exocytosis, and suggest that $VP_2$ represents a vesicle population that has undergone exo-/endocytosis and is able to recycle in the nerve terminal. They are in agreement with the well-documented electrophysiological finding that acetylcholine is released from cholinergic nerve endings in discrete amounts, suggesting that a quantum corresponds to the acetylcholine content of one vesicle.

$VP_1$ represents the mature granule, $VP_2$ the granule after an exo-/endocytotic cycle containing lower amounts of acetylcholine and ATP. Using the monospecific antiserum against purified vesicular proteoglycan, we found, however, a third population of vesicles (Kiene and Stadler, manuscript in preparation). A quantitative immune assay through a vesicle preparation on a zonal gradient gives an additional peak to the lighter region of the sucrose (Fig. 4b).

This population ($VP_0$, Fig. 4b) does not contain acetylcholine and ATP or only very low amounts, and has therefore not been detected previously.

We could confirm this finding by showing that other vesicular markers coincide with the immunochemical marker.

It was, however, not known which state of the vesicle's life cycle this population represented. A possibility was that they were axonally transported vesicles which probably accumulated in the nerve terminal region, but had not yet been filled with transmitter. This possibility was tested by trying to isolate vesicles from axons. As a sensitive vesicle marker we used the in vivo $^{35}S$-sulfate labeling procedure. $^{35}S$-sulfate was injected into the electric lobe and then axons were removed before the peak of radioactivity had reached the electric organ. From the axon material a vesicle isolation procedure was carried out exactly in the same way as from the electric organ. The radioactivity in the zonal eluate which is incorporated into the vesicular proteoglycan exactly coincided with the position of the $VP_0$ fraction (Fig. 4c).

Two conclusions can be drawn from these experiments:

1. Vesicles are transported as preformed particles in the axon.
2. In this cholinergic neuron vesicles are transported as empty or largely empty organelles and are filled with transmitter in the nerve terminal.

## Exocytosis

The proteoglycan antiserum encouraged us to investigate another step in the life cycle of vesicles, namely whether we could monitor vesicular exocytosis in the nerve terminal by immunofluorescence methods. If vesicles fuse with the presynaptic plasma membrane, the proteoglycan inside the vesicles should now be externalized to the synaptic cleft. After electrical stimulation of the electric organ via an electrode implanted into the electric lobe, immunofluorescence in these sections should be much stronger than compared to the resting state because antibodies should in principle only be able to react with externalized and now accessible antigen. Strong increase in immunofluorescence was observed after stimulation providing further evidence for exocytotic release of acetylcholine (Jones et al. 1982).

It was, however, observed that the resting state prior to stimulation already contains significant amounts of externalized antigen. This could be due to the fact that fused vesicle membranes are present to some extent under resting conditions or that proteoglycan has been released exocytotically from vesicles and is present in the synaptic cleft. A recent observation is consistent with the latter view: Basal

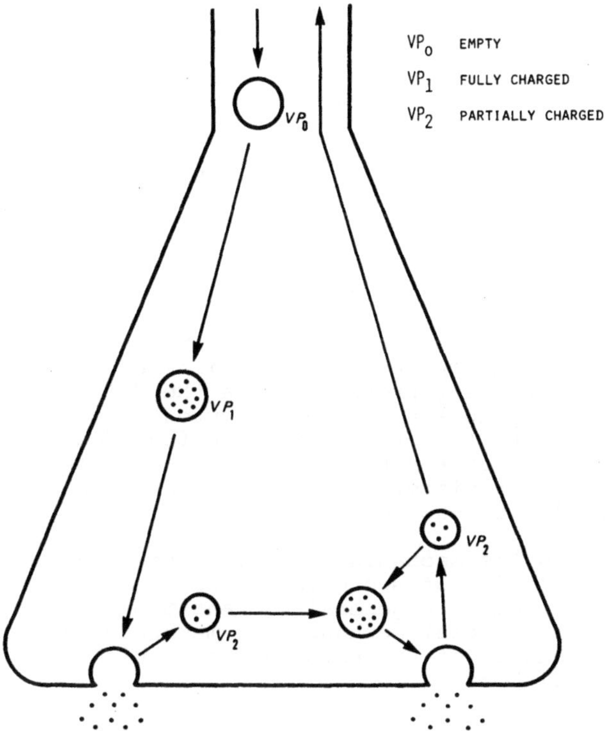


Fig. 5. A model for the life cycle of the cholinergic vesicle in the nerve terminal (explanation see text)

membrane preparations obtained from the electric organ do contain vesicle proteoglycan antigen (Stadler, unpublished). McMahan and coworkers described recently that in the neuromuscular junction and in the electric organ, a factor which probably is a heparansulfate proteoglycan is necessary for clustering of ACh receptors in the synaptic cleft (Godfrey et al. 1984). If this factor is identical to the proteoglycan of cholinergic vesicles that we describe it would constitute the transsynaptic signal needed to keep pre- and postsynaptic elements in alignment. Experiments are in progress now to test this hypothesis.

A heparansulfate proteoglycan is not a unique antigen to cholinergic vesicles from electric organs. The antiproteoglycan antiserum stains using immunofluorescence procedures neuromuscular junctions in mammals (Jones et al. 1982). A heparansulfate proteoglycan was obtained recently from pig brain vesicles. Preliminary results suggest that an antiserum against this proteoglycan might be a marker for cholinergic neurons in the mammalian CNS (Stadler, unpublished).

It is now evident from several systems that a classical transmitter coexists with peptides in the same neuron, although the function of these peptides, whether they act in the sense of a classical transmitter or modulate synaptic transmission, is not clear yet. We have now obtained results in our *Torpedo* model system supporting this concept.

We found that the cholinergic *Torpedo* vesicle contains small amounts of a 7 kd peptide that can be released from the vesicles by hypoosmotic shock under acidic conditions (Stadler and Kiene 1985).

Osmotic shock at neutral pH releases smaller fragments, suggesting that proteolytic cleavage of a precursor is occurring. Since the *Torpedo* synapse is in biochemical terms by far the best-characterized synapse and the only system where pre- and postsynaptic elements can be obtained in high purity, we hope to elucidate aspects of its function.

## The Life Cycle of the Cholinergic Vesicle in the Nerve Terminal

A tentative model of the life cycle of the cholinergic vesicle is presented in Figure 5.

After being manufactured in the Golgi, the synaptic vesicle recognizes the fast transport system in the axon and is carried down to the nerve terminal ($VP_0$ population). In the region of the nerve terminal they are packaged with neurotransmitter and ATP involving carrier-mediated uptake systems, including an ATP-dependent proton pump as the driving force. The mature granule ($VP_1$ population) undergoes exocytosis and releases its soluble contents into the synaptic cleft. The function of ATP is unknown yet. The vesicle is recovered by endocytosis ($VP_2$ population). It can be filled up with transmitter again and is involved with further exo-/endocytotic cycles releasing transmitter.

Altogether the synaptic vesicle and the mechanism of transmitter release appear to be a specialized version of secretion found in many other secretory cells. It differs from this general mechanism by the fast time course of release and the special mechanism of membrane recycling. Probably this is a consequence of the physical separation of the site of secretion from the cell body.

## References


Anderson DC, King SC, Parsons SM (1982) Proton gradient linkage to active uptake of [$^3$H]acetylcholine by Torpedo electric organ synaptic vesicles. Biochemistry 21: 3037-3043

Breer H, Morris SJ, Whittaker VP (1977) Adenosine triphosphatase activity associated with purified cholinergic synaptic vesicles of Torpedo marmorata. Eur J Biochem 80:313-318

Dowdall MJ, Boyne AF, Whittaker VP (1974) ATP: A constituent of cholinergic vesicles. Biochem J 140:1-12

Füldner HH, Stadler H (1982) $^{31}$P-NMR analysis of synaptic vesicles, status of ATP and internal pH. Eur J Biochem 121:519-524

Godfrey EW, Nitkin RM, Wallace BG, Rubin LL, McMahan UJ (1984) Components of Torpedo electric organ and muscle that cause aggregation of acetylcholine receptors on cultures muscle cells. J Cell Biol 99:615-624

Harlos P, Lee DA, Stadler H (1984) Characterization of a $Mg^{2+}$-ATPase and a proton pump in cholinergic synaptic vesicles from the electric organ of Torpedo marmorata. Eur J Biochem 144:441-446

Jones RT, Walker JH, Stadler H, Whittaker VP (1982) Immunohistochemical localization of a synaptic vesicle antigen in a cholinergic neuron under conditions of stimulation and rest. Cell Tissue Res 224:685-688

Luqmani YA (1981) Nucleotide uptake by isolated cholinergic synaptic vesicles: evidence for a carrier of adenosine 5'-triphosphate. Neuroscience 6:1011-1021

Michaelson DM, Ophir I, Angel I (1980) ATP stimulated $Ca^{2+}$ transport into cholinergic Torpedo synaptic vesicles. J Neurochem 35:116-124

Ohsawa K, Dowe GHC, Morris SJ, Whittaker VP (1979) The lipid and protein content of cholinergic synaptic vesicles from the electric organ of Torpedo marmorata purified to constant composition: implications for vesicle structure. Brain Res 161:447-457

Rephaeli A, Parsons SM (1982) Calmodulin stimulation of $^{45}Ca^{2+}$ transport and protein phosphorylation in cholinergic synaptic vesicles. Proc Natl Acad Sci USA 79: 5783-5787

Stadler H, Dowe GHC (1982) Identification of a heparan sulphate-containing proteoglycan as a specific core component of cholinergic synaptic vesicles from Torpedo marmorata. EMBO J 1:1381-1384

Stadler H, Fenwick EM (1983) Cholinergic synaptic vesicles from Torpedo marmorata contain an atractyloside-binding protein related to the mitochondrial ADP/ATP carrier. Eur J Biochem 136:377-382

Stadler H, Füldner HH (1980) Proton NMR detection of acetylcholine status in synaptic vesicles. Nature (London) 286:293-294

Stadler H, Füldner HH (1981) $^{31}$P-NMR analysis of ATP in synaptic vesicles and its relationship to 'in vivo' conditions. Biomed Res 2:673-676

Stadler H, Kiene ML (1985) Costorage of acetylcholine and peptides in synaptic vesicles from Torpedo electric organ. Hoppe-Seyler's Z Biol Chem 366:142

Stadler H, Tashiro T (1979) Isolation of synaptosomal plasma membranes from cholinergic nerve terminals and a comparison of their proteins with those of synaptic vesicles. Eur J Biochem 101:171-178

Stadler H, Tsukita S (1984) Synaptic vesicles contain an ATP-dependent proton pump and show 'knob-like' protrusions on their surface. EMBO J 3:3333-3337

Tashiro T, Stadler H (1978) Chemical composition of cholinergic synaptic vesicles from Torpedo marmorata based on improved purification. Eur J Biochem 90:479-487

Walker JH, Jones RT, Obrocki J, Richardson GP, Stadler H (1982) Presynaptic plasma membranes and synaptic vesicles of cholinergic nerve endings demonstrated by means of specific antisera. Cell Tissue Res 223:101-116

Walker JH, Stadler H, Witzemann V (1984) Calmodulin binding proteins of the cholinergic electromotor synapse: synaptosomes, synaptic vesicles, receptor-enriched membranes, and cytoskeleton. J Neurochem 42:314-320

Whittaker VP, Essman WE, Dowe GHC (1972) The isolation of pure cholinergic synaptic vesicles from the electric organs of elasmobranch fish of the family Torpedinidae. Biochem J 128:833-846

Zimmermann H, Stadler H, Whittaker VP (1981) Structure and function of cholinergic synaptic vesicles. In: Stjarne L, Lagercrantz H, Hedquist P, Wenmalm A (eds) Chemical transmission 75 years. Academic Press, London New York, pp 91-104

# Catecholamine-Storing Vesicles: From Biosynthesis to Exo/Endocytosis

H. Winkler and R. Fischer-Colbrie [1]

## Introduction

Three types of catecholamine-storing vesicles have been characterized: (1) the chromaffin granules from adrenal medulla (2) the large dense core vesicles of sympathetic nerves and (3) the small dense core vesicles, which are also found in these nerves. The composition and molecular organization of chromaffin granules have been analyzed in great detail (see Winkler and Westhead 1980, Winkler and Carmichael 1982). The large dense core vesicles of sympathetic nerve closely resemble chromaffin granules in their biochemical composition and in their functional properties (Klein 1982). In addition it has become obvious that several features of these catecholamine-storing vesicles are of more general significance. As we will discuss below, the peptides present in the content of chromaffin granules, the chromagranins and neuropeptides, have a widespread distribution in endocrine and nervous tissues. Thus studies on their biosynthesis and their function have relevance not only for the adrenal medulla, but, e.g., also for large dense core vesicles in brain containing nonadrenergic transmitters.

The small dense core vesicles of sympathetic nerve contain catecholamines and nucleotides (see Fried 1980), but they do not contain soluble peptides characteristic of chromaffin granules, i.e., chromogranin A and enkephalins (Neuman et al. 1984). On the other hand, their membranes appear to be quite similar, if not identical to those of large dense core vesicles and chromaffin granules. The membranes of small dense core vesicles can also transport catecholamines (Fried 1981) and nucleotides (Aberer et al. 1978) and they contain identical proteins: a $Mg^{2+}$-activated ATPase (Fried 1981), a cytochrome b-561 (Fried 1978) and dopamine β-hydroxylase, although the presence of this latter component is controversial (see Neuman et al. 1984).

The properties of the membranes of adrenergic vesicles are of general significance. The concept that a proton pump creates an electrochemical gradient which in turn drives catecholamin uptake (see Njus 1983) is apparently not confined to adrenergic vesicles. Thrombocyte vesicles take up 5-hydroxytryptamine by the same mechanism (Johnson and Scarpa 1981) and even acetylcholine is transported by an analogous uptake system (Anderson et al. 1982).

In this article we will concentrate on the immunological characterization and on the biosynthesis of the proteins and peptides stored in chromaffin granules. Finally we will discuss that the granule membranes, which transport these secretory components from the Golgi region to the plasma membrane, are used as a recycling container.

[1] Dept. of Pharmacology, University of Innsbruck, Peter Mayr Str. 1a, A-6020 Innsbruck, Austria

36. Colloquium - Mosbach 1985
Neurobiochemistry

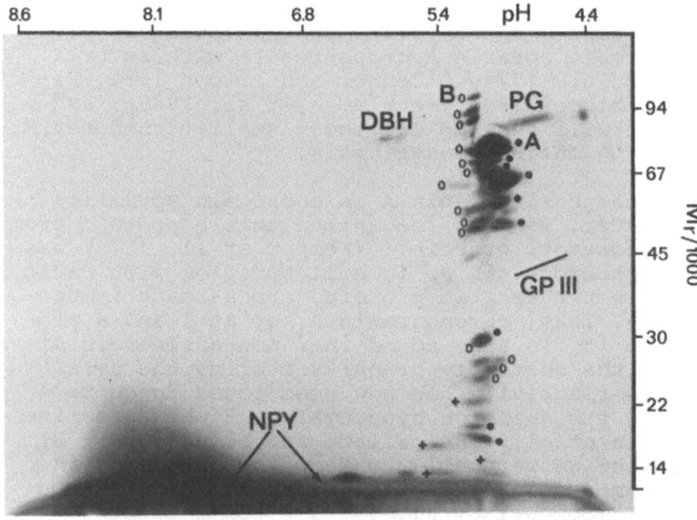


Fig. 1. Characterization of soluble proteins of bovine chromaffin granules by two-dimensional electrophoresis. A photograph of a Coomassie blue-stained pattern is shown. For identifying immunologically related proteins immunoblots were performed (cf. Fischer-Colbrie and Frischenschlager 1985). Antisera against chromogranin A and B, dopamine β-hydroxylase *(DBH)*, glycoprotein III *(GP III*; see Fischer-Colbrie et al. 1984), metenkephalin and neuropeptide Y (NPY) were used. Components reacting with the antiserum against chromogranin A *(A)* are characterized by , those reacting with anti B *(B)* by O, and those reacting with antienkephalin by +. Proteoglycans *(PG)* were identified by their incorporation of $^{35}$S-sulfate and by their sensitivity to chondroitinase digestion (Falkensammer et al. 1985b). Glycoprotein III and neuropeptide Y represent only minor components which cannot be visualized by Coomassie blue staining. The largest component of the chromogranins C migrates just above chromogranin A and is visible as a faint arc in the figure

## Immunological Characterization of the Soluble Proteins and Peptides of Chromaffin Granules

These soluble proteins have been collectively called chromogranins (Blaschko et al. 1967). They constitute, however, a rather complex mixture and therefore a more differentiated nomenclature has to be used. As first shown by Apps (Apps et al. 1980), these proteins can be resolved into at least 30 components by two-dimensional electrophoresis. Minor components have been identified as the enzyme dopamine β-hydroxylase (see Fig. 1) and as the so called glycoprotein III (Fischer-Colbrie et al. 1984). In addition a component having a rather heterogenous pI (see Fig. 1) was recently characterized as a proteoglycan by its ability to incorporate $^{35}$S-sulfate and by its sensitivity to chondroitinase A,B,C (Falkensammer et al. 1985b).

### Chromogranins A

The major component, chromogranin A, which constitutes about 40% of the total proteins, was isolated and characterized some time ago (Smith and Winkler 1967, Kirshner and Kirshner 1969, Helle 1971). The molecular weight of this acidic protein is 75,000; it has properties approaching that of a random coil (Smith and Winkler 1967). Chromogranin A is a glycoprotein (Geissler et al. 1977, Fischer-Colbrie et al. 1982) containing O-glycosidically linked carbohydrate chains (Kiang et al. 1982). An antiserum against chromogranin A cross-reacts with several smaller proteins (Hörtnagl et al. 1974, O'Connor and Frigon 1984, Fischer-Colbrie and Frischenschlager 1985, Settleman et al. 1985), as indicated in Figure 1. As we will show below, these smaller components are formed from chromogranin A during biosynthesis.

Originally it was thought that chromogranin A is a protein specific for adrenergic tissues. However, we then recognized that a protein from parathyroid gland, called secretory protein I (Kemper et al. 1974) was apparently very similar to chromogranin A. In collaboration with David Cohn's group we could show that these two proteins cross-react immunologically (Cohn et al. 1982). Thus, chromogranin A, or at least a protein very closely related to it, is also found in a non-adrenergic tissue, where it is present in the secretory granules storing the parathormone. Chromogranin A-like reactivity has now been found in quite a variety of tissues including the anterior hypophysis and the endocrine pancreas (O'Connor 1983, Cohn et al. 1984, Wilson and Lloyd 1984). By two-dimensional immunoblotting we were able to show that at least in the adenohypophysis the chromogranin A reactive protein had the same properties as chromogranin A (Somogyi et al. 1984, Fischer-Colbrie et al. 1985). This is also true for the chromogranin A-like material found in brain. A thorough immunohistochemical analysis of this organ revealed that chromogranin A has a widespread, but characteristic distribution which is different from that given by various transmitters or neuropeptides (Somogyi et al. 1984).

## Chromogranins B

This family of proteins was described recently (Winkler et al. 1984, Fischer-Colbrie and Frischenschlager 1985, Falkensammer et al. 1985a). They represent the second most abundant proteins in bovine chromaffin granules (see Fig. 1). An antiserum raised against the largest component of the chromogranins B cross-reacts with several smaller proteins, whereas an antiserum raised against highly purified chromogranin A did not cross-react with these proteins. Recent data indicate that the chromogranins B differ also in their sugar composition from the A-family, since only the B-proteins have an affinity for *Pisum sativum* lectin (Apps et al. 1985).

The largest component ($M_r$ 100,000) of the chromogranins B, in contrast to the chromogranin A family, represent only a small portion of all immunologically cross-acting chromogranins B (see Fig. 1).

It is interesting to note that, depending on the species, the relative amounts of the chromogranin A and B vary. In rat chromaffin granules the chromogranins B are the major components (Fischer-Colbrie and Frischenschlager 1985). Chromogranin B is also present in sympathetic nerve, probably in the large dense core vesicles. It is also found in the adenohypophysis, but is absent from neurohypophysis. It is also present in some enterochromaffin cells and in the endocrine pancreas (Fischer-Colbrie et al. 1985). In these tissues the largest chromogranin B component represents the most abundant component of the B-family.

It should be pointed out that great care has to be taken if one wants to obtain specific antisera against chromogranin A or B. Since these two protein families have components comigrating in one-dimensional electrophoresis, it is insufficient to elute the proteins for immunization after such electrophoresis. Antisera reacting with both groups are thus obtained.

## Chromogranins C

This third family of acidic proteins has recently been characterized (Winkler et al. 1986). Its largest component is found just above chromogranin A in 2D-electrophoresis (see Fig. 1).

## Neuropeptides

The report by Schultzberg et al. (1978) that the adrenal medulla contained enkephalin-like material initiated an impressive amount of research. Soon it was established that chromaffin granules contain free enkephalins, but also larger enkephalin-containing peptides were found (see review by Udenfriend and Kilpatrick 1983). When a soluble lysate of bovine chromaffin granules is subjected to two-dimensional immunoblotting with an antiserum against met-enkephalin, three major components are stained specifically (see Fig. 1, Fischer-Colbrie and Frischenschlager 1985). These components are likely to correspond to the enkephalin containing peptides of $M_r$ 18,200, 12,600 and 8,600 (cf. Patey et al. 1984) for one-dimensional electrophoresis). These components, which were isolated by Udenfriend and colleagues, represent the major enkephalin-containing peptides in bovine granules. As shown in Figure 1, they are only minor components of the total soluble proteins. They represent a protein family unrelated to chromogranins A and B.

An antiserum against neuropeptide Y reacts with two peptides moving close to the electrophoresis front. These two peptides represent very minor components (Fischer-Colbrie et al. 1986). An antiserum against bombesin specifically stained the chromogranin B spots. This might indicate that a bombesin-like sequence is found in these proteins (Fischer-Colbrie et al. 1986).

## Biosynthesis of Enkephalins and Chromogranins

The elegant work of several groups has elucidated the biosynthetic pathway of the enkephalins in adrenal medulla (see review by Udenfriend and Kilpatrick 1983). The amino acid sequence of preproenkephalin has been deduced from cDNA by three groups (Gubler et al. 1982, Noda et al. 1982, Comb et al. 1982). Proenkephalin is proteolytically processed within chromaffin granules by proteases. Two of them have been described in some detail. A trypsin-like enzyme is able to split the precursor at paired basic amino acids (Lindberg et al. 1984, Hook 1984). A carboxypeptidase B-like enzyme removes basic amino acids from the carboxyl terminus of peptides, finally yielding free enkephalins (Fricker and Synder 1983). It is interesting to note that this proteolytic processing is apparently occurring at a rather slow rate. Fleminger et al. (1983) calculated that the half life of the precursors is about 1 week. It is therefore not surprising that mature granules, although they no longer contain proenkephalin, still contain significant amounts of large enkephalin peptides ($M_r$ 18,200, 12,600, 8,600) which represent about 50% of the total immunoreactive material. This is obviously quite different from the time course of the conversion of other prohormones to hormones. In the case of insulin, the half life of the prohormone is in the order of 30 min (Steiner et al. 1974). The slow proteolytic processing of the enkephalin precursor is likely to indicate that these precursors have a function of their own (see Udenfriend and Kilpatrick 1983, for further discussion).

This limited proteolytical processing does not only apply to the enkephalin precursors, but also to the major soluble proteins of chromaffin granules, the chromogranins A. In cell-free synthesis chromogranin A is synthesized as two closely related precursors (Kilpatrick et al. 1984, Falkensammer et al. 1985a). In the presence of microsomes (Falken-

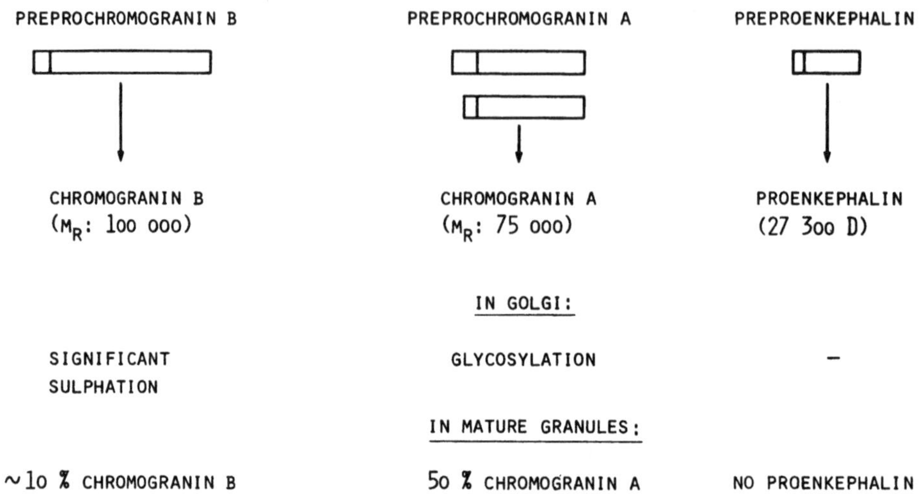


Fig. 2. Cell free and cellular synthesis of chromogranin and enkephalins. These proteins are synthesized as preproteins. After removal of the signal peptides the chromogranins are modified in the Golgi region by sulphation and glycosylation. In chromaffin granules proteolytic processing takes place. The extent of proteolysis varies. In mature granules no proenkephalin is found and only small amounts of chromogranin B are left, whereas chromogranin A is still present as a major component

sammer et al. 1985a), these two prechromogranins A are converted into one component, most likely by the removal of two different signal peptides. For parathyroid secretory protein I, a similar finding has been reported in cell-free synthesis (Majzoub et al. 1982), which is a further indication for the close relationship of these two proteins. A recent claim by Serck-Hanssen and O'Connor (1984) that in cell-free synthesis several precursors are formed for chromogranin A is not supported by the two studies just discussed. It seems likely that their results are explained by an antiserum reacting both with chromogranin A and B, since one of the precursors had a molecular weight of 100,000, which is characteristic of the B-family (see below). During cellular synthesis as shown by pulse chase experiments with $^{3}$H-leucine, prochromogranin A is converted to a slightly larger and slightly more acidic component, apparently, as judged from the time course, by glycosylation in the Golgi complex (Falkensammer et al. 1985a) (Fig. 2). Proteolytic processing of chromogranin A in these exerpiments was slow, since even after a 2-h chase only a very limited breakdown had occurred. This slow processing explains why in mature granules only about 50% of chromogranin A has been broken down (cf. Fig. 1).

In cell-free synthesis, chromogranin B is synthesized as a single component. In the presence of microsomes this prechromogranin B becomes slightly smaller as to be expected from the removal of the signal peptide (Falkensammer et al. 1985a). In cellular synthesis this precursor is converted to a significantly more acidic protein (change in pI from 5.6 to 5.2). This suggested to us that this protein is modified by the incorporation of strong negative charges. Experiments with $^{35}$S-sulfate confirmed this view. This isotope was preferentially incorporated into chromogranin B, chromogranin A became much less labeled (Falkensammer et al. 1985b). Studies by Lee and Huttner (1983) had already shown that in rat PC12 cells sulfate is specifically incorporated into four secretory proteins. Since these PC12 cells are derived from a pheochromocytoma, a tumor of chromaffin tissue, a relationship of these rat proteins to chromogranin B seems quite likely. As already discussed above, in rat chromaffin granules chromogranins B are the major proteins. Furthermore, Hille et al. (1984) have shown that sulfate is incorporated into proteins of several secretory tissues. Chromogranins B were also found to have a widespread distribution in endocrine tissues (see above).

In conclusion, we can state that the major protein families of bovine chromaffin granules, the chromogranins A and B, are synthesized as single proproteins which are then proteolytically processed. This behavior, which they share with the enkephalins and other neuropeptides, and their widespread, but characteristic distribution in many tissues, seem a strong indication that these proteins have a yet undiscovered function after they are secreted.

It has become obvious that the adrenal medulla secretes quite a complex mixture of proteins and peptides. Large dense core vesicles of brain are likely to contain and secrete similar "cocktails". We may soon have complete data on these secretory products, but we can expect that it will take some time to unravel the functional contribution made by each component.

## Membranes of Chromaffin Granules – Recyling Containers?

All these secretory products we have just discussed are secreted from the chromaffin cell by exocytosis. During this process the membranes of chromaffin granules become incorporated into the plasma membranes. It has long been recognized that some membrane retrieval must exist in order to prevent a permanent enlargement of the cell surface during stimulation (Douglas 1968, Holtzman and Dominitz 1968).

Recent studies have enabled us to answer more specific questions in this context. (1) How specific and how fast is this membrane retrieval? (2) What is the final fate of the retrieved membranes? The use of antibodies raised against antigens on the inner surface of chromaffin granules was crucial for these studies. During stimulation such antigens of chromaffin granules become exposed on the cell surface of isolated chromaffin cells (Wildman et al. 1981, Phillips et al. 1983, Patzak et al. 1984). There are two possibilities to retrieve these membranes (see Fig. 3): (1) unspecific removal to prevent enlargement of the cell (2) specific retrieval removing the granule membranes and their antigens. A prerequisite for specific removal is that the granule antigens do not spread translationally and stay together. Apparently this is happening, since even in prefixed cells the membrane antigens exposed during stimulation are present as discrete patches (Patzak et al. 1984). In order to get an idea on the time course of this, retrieval quantitative data had to be obtained. Two methods proved to be useful (1) a cytotoxicity test (Lingg et al. 1983) and (2) the use of a fluorescence-activated cell sorter for evaluating the degree of immunofluorescence (Patzak et al. 1984). Both methods gave the same results. It takes about 30 min, to remove granule membrane patches from the cell surface which had been exposed during a stimulation releasing about 15% of the total catecholamine pool (Patzak et al. 1984). This overall retrieval, of course, does not tell us anything about the single event of membrane retrieval. This has recently been studied by an elegant electrophysiological method, the patch clamp technique (Hamill et al. 1981). Insertion and removal of granule membranes is picked up in the form of minute changes in the cell membrane capacitance (Neher and Marty 1982). These fusion and fission events apparently occurred constantly and independently of each other.

In 1972 we described that in adrenal medulla the synthesis rate of the granule membrane proteins is lower than that of the soluble ones

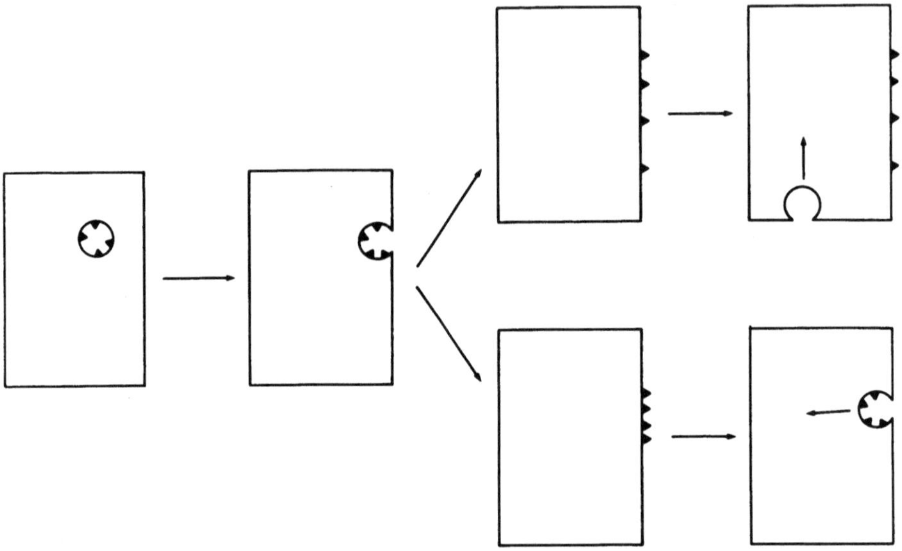

Fig. 3. Exo-/endocytosis cycle. Two possible mechanisms are illustrated. After exocytosis the antigens of chromaffin granules *(triangles)* spread translationally in the plasma membranes. Removal of membranes to prevent enlargement does not involve specific membrane retrieval. Chromaffin granule membranes are specifically retrieved. This mechanism has recently been supported by experiments using specific antibodies against the granule membrane antigens

(Winkler et al. 1972). Based on this finding we postulated that the membranes of chromaffin granules are re-used for several secretory cycles. Further evidence for this was obtained by using exogenous tracers like cationized ferritin, which during stimulation were taken up into secretory cells via the retrieved membranes, and were finally found in the Golgi region and in newly formed secretory vesicles (see Herzog 1981, Suchard et al. 1981). These studies demonstrated recycling of membranes, but the specific nature of the recycled membranes could not be established. We have therefore used antibodies against chromaffin granule membranes for this purpose. Isolated chromaffin cells were first stimulated, then treated with antibodies against glycoprotein III followed by anti IgG labeled with colloidal gold. After this treatment, cells were incubated again. The specific gold label was first found in patches on the plasma membrane, which were selectively retrieved by coated vesicles. Label also appeared in vesicles close to the Golgi region and finally in what appeared to be new chromaffin granules (Patzak and Winkler 1986). Thus it appears likely that the specialized membranes of chromaffin granules are used as a recycling container for a shuttle service from the Golgi region to the plasma membrane.

## Conclusion

Immunological studies of the soluble proteins of chromaffin granules have revealed that most of these proteins belong to three families, i.e., chromogranins A, B, C, and enkephalin-containing peptides. These families are derived by proteolytic processing from single proproteins. All three protein groups have a widespread distribution in endocrine and nervous tissue. The membranes of chromaffin granules which are incorporated into the plasma membrane during exocytosis are specifically retrieved and probably used as a shuttle service from the Golgi region to the plasma membrane.

*Acknowledgments.* Our work reported here was supported by the Dr. Legerlotz-Stiftung.

## References


Aberer W, Stitzel R, Winkler H, Huber E (1979) Accumulation of $^3$H-ATP in small dense core vesicles of superfused vasa deferentia. J Neurochem 33:797-801

Anderson DC, King SC, Parsons SM (1982) Proton gradient linkage to active uptake of $^3$H acetylcholine by Torpedo electric organ synaptic vesicles. Biochemistry 21: 3037-3043

Apps DK, Pryde JG, Phillips JH (1980) Cytochrome b-561 is identical with chromomembrin B, a major polypeptide of chromaffin granule membranes. Neuroscience 5: 2279-2287

Apps DK, Phillips JH, Purves FC (1985) Glycoproteins of the chromaffin-granule matrix. Neuroscience 16:477-487

Blaschko H, Comline RS, Schneider FH, Silver M, Smith AD (1967) Secretion of a chromaffin granule protein, chromogranin from the adrenal gland after splanchnic stimulation. Nature (London) 215:58-59

Cohn DV, Zangerle R, Fischer-Colbrie R, Chu LLH, Elting JJ, Hamilton JW, Winkler H (1982) Similarity of secretory protein I from parathyroid gland to chromogranin A from adrenal medulla. Proc Natl Acad Sci USA 79:6056-6059

Cohn DV, Elting JJ, Frick M, Elde R (1984) Selective localization of the parathyroid secretory protein-I/adrenal medulla chromogranin A protein family in a wide variety of endocrine cells of the rat. Endocrinology 114:1963-1974

Comb M, Seeburg PH, Adelman J, Eiden L, Herbert E (1982) Primary structure of the human Met- and Leu-enkephalin precursor and its mRNA. Nature (London) 295:663-666

Douglas WW (1986) Stimulus-secretion coupling: the concept and clues from chromaffin and other cells. Br J Pharmacol Chemother 34:451-474

Dowd DJ, Edwards C, Englert D, Mazurkiewicz JE, Ye HZ (1983) Immunofluorescent evidence for exocytosis and internalization of secretory granule membrane in isolated chromaffin cells. Neuroscience 10:1025-1033

Falkensammer G, Fischer-Colbrie R, Richter K, Winkler H (1985a) Cell-free and cellular synthesis of chromogranin A and B of bovine adrenal medulla. Neuroscience 14:735-746

Falkensammer G, Fischer-Colbrie R, Winkler H (1985b) Biogenesis of chromaffin granules: Incorporation of sulphate into chromogranin B and into a proteoglycan. J Neurochem 45:1475-1480

Fischer-Colbrie R, Diez-Guerra J, Emson PC, Winkler H (1986) Bovine chromaffin granules: Immunological studies with antisera against neuropeptide $\gamma$, met-enkephalin and bombesin. Neuroscience, in press

Fischer-Colbrie R, Frischenlager I (1985) Immunological characterization of secretory proteins of chromaffin granules: Chromogranins A, chromogranins B and enkephalin-containing peptides. J Neurochem 44:1854-1861

Fischer-Colbrie R, Lassmann H, Hagn C, Winkler H (1985) Immunological studies on the distribution of chromogranin A and B in endocrine and nervous tissues. Neuroscience, in press

Fischer-Colbrie R, Schachinger M, Zangerle R, Winkler H (1982) Dopamine $\beta$-hydroxylase and other glycoproteins from the soluble content and the membranes of adrenal chromaffin granules: Isolation and carbohydrate analysis. J Neurochem 38:725-732

Fischer-Colbrie R, Zangerle R, Frischenschlager I, Weber A, Winkler H (1984) Isolation and immunological characterization of a glycoprotein from adrenal chromaffin granules. J Neurochem 42:1008-1016

Fleminger G, Ezra E, Kilpatrick DL, Udenfriend S (1983) Processing of enkephalin-containing peptides in isolated bovine adrenal chromaffin granules. Proc Natl Acad Sci USA 80:6418-6421

Fricker LD, Snyder SH (1983) Purification and characterization of enkephalin convertase, an enkephalin-synthesizing carboxypeptidase. J Biol Chem 258:10950-10955
Fried G (1978) Cytochrome b-561 in sympathetic nerve terminal vesicles from rat vas deferens. Biochim Biophys Acta 507:175-177
Fried G (1980) Small noradrenergic storage vesicles isolated from rat vas deferens-biochemical and morphological characterization. Acta Physiol Scand Suppl 493:1-28
Fried G (1981) Noradrenaline release and uptake in isolated small dense cored vesicles from rat seminal ducts. Acta Physiol Scand 112:41-46
Geissler D, Martinek A, Margolis RU, Margolis RK, Skrivanek JA, Ledeen R, König P, Winkler H (1977) Composition and biogenesis of complex carbohydrates of ox adrenal chromaffin granules. Neuroscience 2:685-693
Gubler U, Seeburg P, Hoffman BJ, Gage LP, Udenfriend S (1982) Molecular cloning establishes proenkephalin as precursor of enkephalin-containing peptides. Nature (London) 295:206-208
Hamill OP, Marty A, Neher E, Sakmann B, Sigworth FJ (1981) Improved patch clamp techniques for high-resolution current recording from cells and cell-free membrane patches. Pfluegers Arch 391:85-100
Helle KB (1971) Biochemical studies of the chromaffin granule-II. Properties of membrane-bound and water-soluble forms of chromogranin A and dopamine β-hydroxylase activity. Biochim Biophys Acta 245:94-104
Herzog V (1981) Endocytosis in secretory cells. Philos Trans R Soc London Ser B 296: 67-72
Hille A, Rosa P, Huttner WB (1984) Tyrosine sulfation: a post-translational modification of proteins destined for secretion? FEBS Lett 177:129-134
Hörtnagl H, Lochs H, Winkler H (1974) Immunological studies on the acidic chromogranins and on dopamine β-hydroxylase (E.C.1.14.2.1) of bovine chromaffin granules. J Neurochem 22:197-199
Holtzman E, Dominitz R (1968) Cytochemical studies of lysosomes, golgi apparatus and endoplasmatic reticulum in secretion and protein uptake by adrenal medulla cells of the rat. J Histochem Cytochem 16:320-336
Hook VYH (1984) Carboxypeptidase B-like activity for the processing of enkephalin precursors in the membrane component of bovine and adrenomedullary chromaffin granules. Neuropeptides 4:117-126
Johnson RG, Scarpa A (1981) The electron transport chain of serotonin-dense granules of platelets. J Biol Chem 256:11966-11969
Kemper B, Habener JF, Rich A, Potts Jr JT (1974) Parathyroid secretion: discovery of a major calcium-dependent protein. Science 184:167-169
Kiang W-L, Krusius T, Finne J, Margolis RU, Margolis RK (1982) Glycoproteins and proteoglycans of the chromaffin granule matrix. J Biol Chem 257:1651-1659
Kilpatrick L, Gavine F, Apps D, Phillips J (1983) Biosynthetic relationship between the major matrix proteins of adrenal chromaffin granules. FEBS Lett 164:383-388
Kirshner AG, Kirshner N (1969) A specific soluble protein from the catecholamine storage vesicles of bovine adrenal medulla. Biochim Biophys Acta 181:219-225
Klein RH (1982) Chemical composition of the large noradrenergic vesicles. In: Klein RL, Lagercrantz H, Zimmermann H (eds) Neurotransmitter vesicles. Academic Press, London New York, pp 133-173
Lee RWH, Huttner WB (1983) Tyrosine-O-sulfated proteins of PC12 pheochromocytoma cells and their sulfation by a tyrosylprotein sulfotransferase. J Biol Chem 258: 11326-11334
Lindberg I, Yang H-YT, Costa E (1984) Further characterization of an enkephalin-generating enzyme from adrenal medullary chromaffin granules. J Neurochem 42: 1411-1419
Lingg G, Fischer-Colbrie R, Schmidt W, Winkler H (1983) Exposure of an antigen of chromaffin granules on cell surface during exocytosis. Nature (London) 301:610-611
Majzoub JA, Dee PC, Habener JF (1982) Cellular and cell-free processing of parathyroid secretory proteins. J Biol Chem 257:3581-3588
Neher E, Marty A (1982) Discrete changes of cell membrane capacitance observed under conditions of enhanced secretion in bovine adrenal chromaffin cells. Proc Natl Acad Sci USA 79:6712-6716
Neuman B, Wiedermann CJ, Fischer-Colbrie R, Schober M, Sperk G, Winkler H (1984) Biochemical and functional properties of large and small-dense core vesicles in sympathetic nerves of rat and ox vas deferens. Neuroscience 13:921-931

Njus D (1983) The chromaffin vesicle and the energetics of storage organelles. J Auton Nerv Syst 7:35-40
Noda M, Furutani Y, Takahashi H, Toyosato M, Hirose T, Inayama S Nakanishi S, Numa S (1982) Cloning and sequence analysis of cDNA for bovine adrenal preproenkephalin. Nature (London) 295:202-206
O'Connor DT (1983) Chromogranin: widespread immunoreactivity in polypeptide hormone producing tissues and in serum. Regul Peptides 6:263-280
O'Connor DT, Frigon RP (1984) Chromogranin A, the major catecholamine storage vesicle soluble protein. J Biol Chem 259:3237-3247
Patey G, Liston D, Rossier J (1984) Characterization of new enkephalin-containing peptides in the adrenal medulla by immunoblotting. FEBS Lett 172:303-308
Patzak A, Böck G, Fischer-Corbie R, Schauenstein K, Schmidt W, Lingg G, Winkler H (1984) Exocytotic exposure and retrieval of membrane antigens of chromaffin granules: Quantitative evaluation of immunofluorescence on the surface of chromaffin cells. J Cell Biol 98:1817-1824
Patzak A, Winkler H (1986) Exocytotic exposure and recycling of membrane antigens of chromaffin granules: Ultrastructural evaluation after immunolabeling. J Cell Biol in press
Phillips JH, Burridge K, Wilson SP, Kirshner N (1983) Visualization of the exocytosis/ endocytosis secretory cycle in cultured adrenal chromaffin cells. J Cell Biol 97:1906-1917
Schultzberg M, Lundberg JM, Hökfelt T, Terenius L, Brandt J, Elde RP, Goldstein M (1978) Enkephalin like immunoreactivity in gland cells and nerve terminals of the adrenal medulla. Neuroscience 3:1169-1186
Serck-Hanssen G, O'Connor DT (1984) Immunological identification and characterization of chromogranins coded by Poly(A) mRNA from bovine adrenal medulla and pituitary gland and human phaeochromocytoma. J Biol Chem 259:11597-11600
Settleman J, Fonseca R, Nolan J, Hogue-Angeletti R (1985) Relationship of multiple forms of chromogranin. J Biol Chem 260:1645-1651
Smith AD, Winkler H (1967) Purification and properties of an acidic protein from chromaffin granules of bovine adrenal medulla. Biochem J 103:483-492
Somogyi P, Hodgson AJ, De Potter RW, Fischer-Colbrie R, Schober M, Winkler H, Chubb IW (1984) Chromogranin immunoreactivity in the central nervous system. Immunochemical characterization, distribution and relationship to catecholamine and enkephalin pathways. Brain Res Rev 8:193-230
Steiner DF, Kemmler W, Tager HS, Peterson JD (1974) Proteolytic processing in the biosynthesis of insulin and other proteins. Fed Proc 33:2105-2115
Suchard SJ, Corcoran JJ, Pressman BC, Rubin RW (1981) Evidence for secretory granule membrane recycling in cultured adrenal chromaffin cells. Cell Biol Int Rep 5: 953-962
Udenfriend S, Kilpatrick DL (1983) Biochemistry of the enkephalins and enkephalin-containing peptides. Arch Biochem Biophys 221:309-323
Wildman J, Dewair M, Matthaei H (1981) Immunochemical evidence for exocytosis in isolated chromaffin cells after stimulation with depolarizing agents. J Neuroimmunol 1:353-364
Wilson BS, Lloyd RV (1984) Detection of chromogranin in neuroendocrine cells with a monoclonal antibody. Am J Pathol 115:458-468
Winkler H, Apps DK, Fischer-Colbrie R (1986) The molecular function of adrenal chromaffin granules. Neuroscience, in press
Winkler H, Carmichael SW (1982) The chromaffin granule. In: Poisner AM, Trifaro JM (eds) The secretory granule. Elsevier Biomed Press, Amsterdam, pp 3-79
Winkler H, Westhead E (1980) The molecular organization of adrenal chromaffin granules. Neuroscience 5:1803-1823
Winkler H, Schöpf JAL, Hörtnagl H, Hörtnagl H (1972) Bovine adrenal medulla: subcellular distribution of newly synthesized catecholamines, nucleotides and chromogranins. Naunyn-Schmiedebergs Arch Exp Pathol Pharmakol 273:43-61
Winkler H, Falkensammer G, Patzak A, Fischer-Colbrie R, Schober M, Weber A (1984) Life cycle of the catecholaminergic vesicles: From biogenesis to secretion. In: Vizi ES, Magyar K (eds) Regulation of transmitter function: Basic and clinical aspects. Akadêmiai Kiadô, Budapest, pp 65-73

# Activation and Deactivation of the Cyclic Nucleotide Enzyme Cascade in Visual Rod Cells

H. Kühn, S. W. Hall, M. Wehner, and U. Wilden[1]

## Introduction

Absorption of light by rhodopsin in vertebrate rod cell outer segments (ROS) initiates a chain of events that finally lead to a decrease in ion permeability of the plasma membrane and to hyperpolarization of the rod cell. Most of the rhodopsin is located in the stack of disk membranes, physically separated from the surrounding plasma membrane. It is generally believed that light-induced changes in the concentration of a cytoplasmic diffusible substance ("internal messenger") mediate between light absorption at the disk membrane and closure of ion channels at the plasma membrane. Cyclic 3',5'-guanosine monophosphate (cGMP) and $Ca^{2+}$ ions are the most favored candidates. Both of them affect the light-sensitive ion permeability: cGMP leads to an increase, $Ca^{2+}$ to a decrease of the dark inward current (cf. Miller 1981, Fesenko et al. 1985).

The present paper is concerned with early events occurring at the disk membrane after photoexcitation of rhodopsin. It will be shown that rhodopsin communicates the message of photon absorption to the rod cell by specifically interacting with several ROS proteins; a sequence of protein-protein interactions leads first to the activation, and at a later stage to the deactivation, of the enzyme cascade that catalyzes hydrolysis of cGMP.

Experiments were performed with isolated ROS fragments purified from bovine retina. These preparations contain the disk membranes, disrupted plasma membrane, peripherally associated enzymes, and a major part of the cytoplasmic (soluble) proteins (Kühn 1981).

## Light-Induced Conformational Changes of Rhodopsin

Light absorption by rhodopsin causes rapid isomerization of its chromophore, 11-cis retinal, to the all-trans configuration. The early, high-energy photoproduct (bathorhodopsin) relaxes through a sequence of spectrally defined photointermediates (lumirhodopsin, metarhodopsins I, II, and III) until the chromophore finally dissociates from the apoprotein, opsin (Wald 1968). Each of these intermediates reflects a distinct protein conformational state in the hydrophobic, chromophore-binding core of rhodopsin. Conformational changes at rhodopsin's cytoplasmic-exposed

---

*Abbreviations:* ROS, rod outer segments; PDE, phosphodiesterase; cGMP, cyclic 3',5'-guanosine monophosphate; G-protein, guanosine nucleotide binding protein with subunits $G_\alpha$, $G_\beta$ and $G_\gamma$.

[1]Institut für Neurobiologie der KFA Jülich GmbH, Postfach 1913, D-5170 Jülich, FRG

36. Colloquium - Mosbach 1985
Neurobiochemistry

surface, on the other hand, have been more difficult to demonstrate and seem to be very locally restricted. One SH group (Chen and Hubbel 1978) and two methionine residues (Pellicone et al. 1985) become more accessible to chemical modification, and a peptide bond near the carboxyl terminus becomes more accessible to proteolytic cleavage following illumination (Kühn et al. 1982).

The most significant indications of surface-conformational changes, however, are provided by the specific, light-induced interactions of rhodopsin with several peripheral and cytoplasmic proteins of ROS, as described in the following sections.

## Light-Dependent Phosphorylation of Rhodopsin

Dark-adapted rhodopsin is unphosphorylated. Photolyzed rhodopsin becomes accessible to a specific protein kinase, which, independently of cyclic nucleotides or $Ca^{2+}$, incorporates up to nine phosphate groups into serine and threonine residues of the protein (Kühn and Dreyer 1972, Bownds et al. 1972, Wilden and Kühn 1982). ATP is the preferred phosphate donor. Phosphorylation of photolyzed rhodopsin is 50 - 100 times faster than that of dark-adapted rhodopsin, indicating a highly specific recognition by the kinase of a light-induced conformational change at rhodopsin's cytoplasmic surface. To our knowledge, rhodopsin is the first receptor protein whose phosphorylation has been demonstrated, but only very recently has a function for this phosphorylation reaction been established (see later).

## Light-Induced Interactions of Proteins with Rhodopsin: Sedimentation Studies

Rhodopsin is the predominant integral membrane protein of the disks, constituting more than 90% of the total protein in thoroughly washed disk membranes. When ROS fragments are centrifuged in physiological ionic strength buffer, the membranes containing rhodopsin and the "peripherally associated proteins" sediment, whereas a few dozen proteins remain in the supernatant. Two of these "soluble proteins" are only soluble in dark-adapted preparations but become membrane-associated upon illumination (Kühn 1978): the enzyme *rhodopsin kinase* ($M_r$ 68,000) and a protein of $M_r$ 48,000 *(48 kd-protein)*. Both proteins consist of a single polypeptide (Table 1).

Their association with bleached disk membranes is spontaneously but slowly reversed: about 1 h after illumination both proteins are soluble again. This reversible binding has been used to separate the two light-dependent proteins from the bulk of soluble ROS proteins, yielding a preparation consisting of about 95% pure 48 kd-protein and 2 - 3% kinase (Kühn 1978).

The 48 kd-protein has recently been shown to be identical with the so-called S-antigen, a soluble ROS protein that can cause an autoimmune disease of the eye (autoimmune uveoretinitis) (Pfister et al. 1985). We have since used published procedures for purification of S-antigen (Dorey et al. 1982) to obtain 48 kd-protein preparations free of kinase activity.

Table 1. Proteins that undergo light-induced binding to the disk membrane[a]

| Name (synonym) | Native protein[b] ($M_r$) | Poly-peptides ($M_r$) | Solubility at physiological ionic strength in darkness | Effects of nucleotides |
|---|---|---|---|---|
| Rhodopsin-kinase (opsin-kinase) | 69,000 | 68,000 | Mostly soluble | ATP increases solubility |
| 48 kd-Protein | 48,000 | 48,000 | Soluble | ATP and GTP strongly enhance light-induced binding |
| G-Protein (GTP-binding protein, GTPase, transducin) | 85,000 | 40,000 (α)<br>35,000 (β)<br>8,000 (γ) | Membrane-associated[c] | GTP and analogs dissociate $G_\alpha$-GTP from R* and $G_{\beta\gamma}$ |

[a]This table summarizes results from Kühn (1978, 1980, 1984). Either bleached, or unbleached, suspensions of ROS membranes were centrifuged. The clear supernatants were assayed: for rhodopsin kinase activity using kinase-free, illuminated rhodopsin-containing membranes as a substrate; for G-protein by measuring its R* dependent GTPase activity (Kühn 1980); and for the presence of polypeptides using SDS-PAGE

[b]Determined from calibrated Sephadex G100 columns

[c]Illumination at physiological ionic strength changes the mode of association of G-protein with the membrane. This is revealed by subsequently treating the illuminated membranes with low ionic strength buffer: the G-protein is readily extractable, at low ionic strength, from dark-kept but not from previously illuminated membranes (see also Fig. 1)

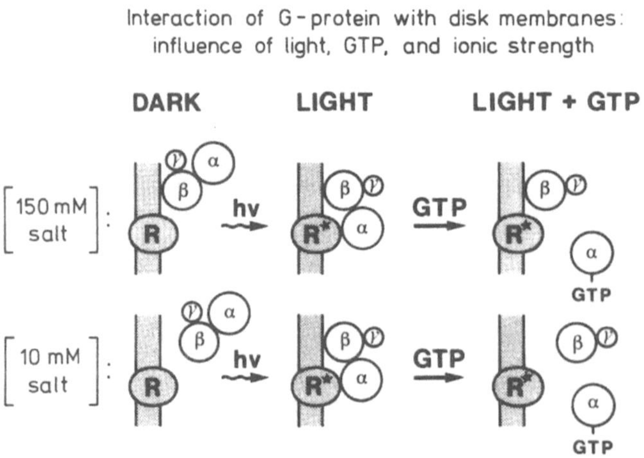


Fig. 1. Scheme, resulting from sedimentation experiments, of interactions between G-protein and the disk membrane. At physiologic ionic strength *(upper row)*, G-protein (subunits $G_\alpha$, $G_\beta$, and $G_\gamma$) is membrane-associated both in the dark and in the light. The difference between dark association and light-induced binding is revealed by treatment with low ionic strength buffer *(lower row)*. Light plus GTP at physiological ionic strength leads to preferential solubilization of $G_\alpha$ in its GTP-binding form ($G_\alpha$-GTP), whereas at low ionic strength all three subunits are eluted

*G-protein* and *phosphodiesterase* (PDE), the two major *peripherally associated* proteins of disks, are membrane-associated at physiological ionic strength (salt concentrations between 100 and 200 mM) and both become eluted at low ionic strength (< 20 mM salt) from dark-adapted ROS. After illumination, on the other hand, only the PDE is eluted, whereas the G-protein becomes membrane-bound (Fig. 1; Kühn 1980). This "light induced binding" of G-protein occurs independently of the ionic strength at which the membranes are illuminated, but is best revealed at low ionic strength when the protein is readily extractable from dark-kept but not from illuminated membranes.

The G-protein is specifically and rapidly solubilized from bleached disks by GTP (Kühn 1980) or poorly hydrolyzable analogs of GTP (GTP$\gamma$S, GppNHp). Experiments with $^3$H and $\gamma$-$^{32}$P-labeled GTP have shown that this solubilization is caused by the exchange of GTP for previously bound GDP on the G-protein: in the GDP-binding form, the protein binds to bleached membranes, and in the GTP-binding form it is soluble (Kühn 1980, 1981, 1984, Fung and Stryer 1980, Fung 1983). At low ionic strength, all three subunits of G-protein are eluted by GTP (Fig. 1, lower part; Kühn 1980); at moderate ionic strength, $G_\alpha$-GTP is preferentially eluted, most of $G_{\beta\gamma}$ remaining membrane-associated (Fig. 1, upper part; Kühn 1981). The nucleotide binding site resides on $G_\alpha$ (Fung et al. 1981), whereas $G_{\beta\gamma}$ seems to anchor $G_\alpha$-GDP to the membrane. $G_\alpha$ possesses intrinsic GTPase activity; after hydrolysis of the bound GTP, $G_\alpha$-GDP reassociates to $G_{\beta\gamma}$ and to the membrane at moderate ionic strength (see also Fig. 6).

The light-induced binding of G-protein and its subsequent specific elution by GTP can be used for convenient purification of G-protein. Semipurified PDE (about 90% pure) is obtained as a by-product (Kühn 1982). Rod outer segments contain roughly one molecule of PDE per 10 G-proteins and 100 rhodopsins.

The sedimentation studies summarized here are consistent with the idea, but do not unequivocally prove, that the three light-dependent proteins (see Table 1) bind directly to photoexcited rhodopsin (R*). For the kinase, this assumption is plausible, since R* is its substrate. For G-protein and 48 kd-protein, more rigorous tests are required and have been provided (see below).

## Activation of the cGMP-Degrading Enzyme Cascade

The light-induced binding of G-protein, and its GTP-induced dissociation, are the first steps in a light-activated enzyme cascade that leads to amplified activation of cGMP-phosphodiesterase (Fig. 2). Several laboratories have been involved in elucidating this cascade. The G-protein was discovered by Wheeler and Bitensky (1977) from its intrinsic GTPase activity and was proposed to serve as a mediator in PDE activation, analogous to the system of hormone-triggered adenylate cyclase activation in which related G-proteins mediate between hormone receptor and cyclase. The $\alpha$-subunit of G-protein in the GTP-binding form ($G_\alpha$-GTP) activates the PDE (Fung et al. 1981, Uchida et al. 1981) by releasing a 13 kd inhibitory subunit of PDE (Hurley and Stryer 1982).

The most noteworthy aspect of this cascade is its high degree of amplification: one photoexcited rhodopsin molecule (R*) can trigger the activation of hundreds of molecules of G-protein (Fung and Stryer 1980) and, therefore, of PDE (Yee and Liebman 1978). Since the fully acti-

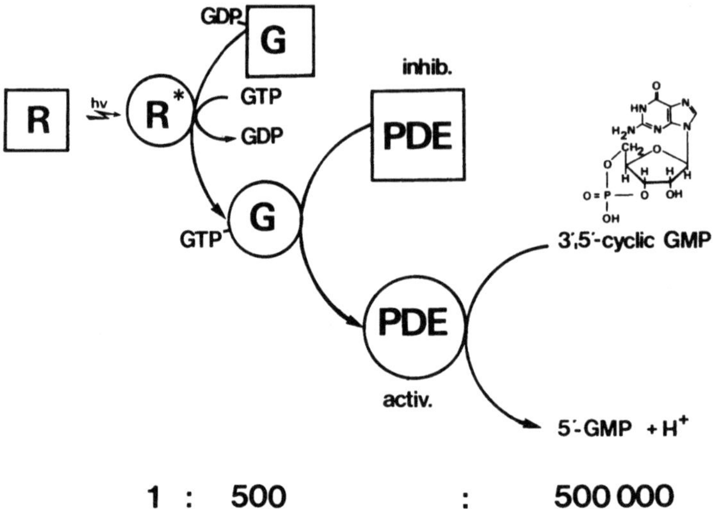


Fig. 2. Scheme of the light-activated cGMP enzyme cascade. Photoexcited rhodopsin *(R*)* catalyzes, by direct molecular contact with G-protein *(G)*, the exchange of GTP for bound GDP on G-protein. The active (GTP-binding) form of G-protein then activates the cGMP-phosphodiesterase *(PDE)* by releasing an inhibitory constraint from the PDE. The light signal is amplified in two stages *(numbers at bottom)*: absorption of a single photon can lead to the activation of several hundred enzyme molecules and to the degradation of 500,000 substrate molecules per second

vated PDE has a turnover number of 2000 $s^{-1}$ (Baehr et al. 1979), bleaching of rhodopsin results in a two-step amplification cascade leading to the hydrolysis of hundreds of thousands of molecules of cGMP per second per absorbed photon.

Evidence that R* directly triggers this cascade, through a specific molecular interaction with G-protein, comes from several experimental approaches (Kühn 1984).

1. The membrane sedimentation studies summarized above (see Fig. 1 and Table 1), showing light-induced binding and GTP-induced dissociation of G-protein, suggest an interaction of R* with G-protein but do not strictly rule out the involvement of other factors such as light-induced changes in membrane potential etc.
2. The most compelling evidence comes from binding studies in a membrane-free system, using purified proteins in detergent solution. Purified, lipid-free rhodopsin, anchored via its carbohydrate side chains to Concanavalin A-Sepharose, binds purified G-protein only after illumination, and the G-protein bound to R* is released by GTP or GTPγS (Kühn 1984).
3. Proteolytic removal of a few amino acids from a hydrophilic "loop" region connecting helices V and VI at rhodopsin's cytoplasmic surface abolishes rhodopsin's capacity to catalyse GDP/GTP exchange on G-protein (Kühn and Hargrave 1981, Wehner et al., in preparation). Rhodopsin's site of interaction with G-protein is distinct from the bulk of phosphorylation sites.
4. Various spectroscopic studies have shown that the binding of G-protein to R* shifts the equilibria existing between the photointermediates metarhodopsin I, II, and III, toward metarhodopsin II, at the cost of both I and III (Emeis et al. 1982, Pfister et al.

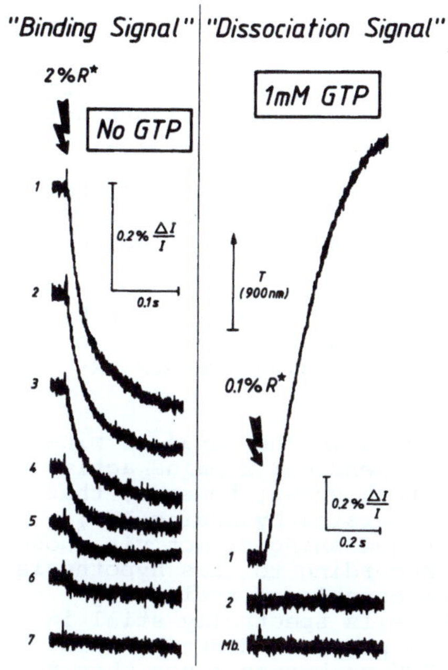


Fig. 3. Flash-induced changes in transmittance ($T_{900\ nm}$) of ROS suspensions at 20°C. *Left side* "Binding signal" obtained in absence of GTP. Seven successive flashes, each bleaching 2% of the rhodopsin in the same ROS suspension. The signal saturates at about 10% total bleaching extent; since ROS contain about 1 molecule of G-protein per 10 rhodopsins, this saturation behavior reflects a 1 : 1 stoichiometric binding of G-protein to R*. *Right side* "Dissociation signal" obtained from sonicated ROS membrane vesicles in presence of 1 mM GTP. A single flash producing only 0.1% R* gives rise to a saturated signal, reflecting the complete turnover of G-protein into the GTP-binding form (100 G-proteins turned over by 1 R* at this light intensity). A second flash *(2)* produces no further signal. Hypoosmotically washed disk membranes *(Mb.)* give neither a binding nor a dissociation signal; both signals can be reconstituted by adding purified G-protein. (Kühn et al. 1981)

1983). This indicates specific coupling between metarhodopsin II and G-protein. It also suggests that metarhodopsin II, which is formed in vivo within about one millisecond after light absorption, is the active photointermediate (R*) that triggers the enzyme cascade.

## Kinetics of Interactions Between R* and G-Protein

The binding of G-protein to R* in the absence of GTP, and its dissociation from R* after nucleotide exchange in the presence of GTP, give rise to rapid and large changes in the light-scattering of rod outer segment membrane suspensions (Fig. 3), termed binding signal and dissociation signal, respectively (Kühn et al. 1981). Both signals occur in the time range of 10 - 100 milliseconds. The *binding signal* is seen as a flash-light induced increase in turbidity; it saturates when an amount of rhodopsin equimolar to the amount of G-protein present has been bleached, reflecting a 1 : 1 stoichiometric binding of G-protein to R* in the absence of GTP.

The *dissociation signal*, a flash-induced decrease in turbidity observed in the presence of GTP, saturates at much lower light intensities, reflecting the highly amplified turnover of all of the G-protein molecules by relatively few R* molecules. Similar amplification numbers of 100 - 500 molecules of G-protein, activated by a single R* molecule, are obtained from the saturation behavior of the dissociation signal (Fig. 3, right side; Kühn et al. 1981) and from the binding of radiolabeled GTP to G-protein (Fung and Stryer 1980, Wehner et al., in preparation). The dissociation signal most probably reflects the dissociation of $G_{\alpha}$-GTP from the membranes into solution (Kühn 1984), whereas the binding signals reflects more complicated structural events triggered by the binding of G-protein to R* (Hofmann et al. 1981).

## Deactivation of the Enzyme Cascade

The rod cell responds to a light stimulus only transiently and returns to baseline within a few seconds after a weak flash. If the light-activated PDE enzyme cascade triggers excitation of the rod cell, it should be deactivated in the same time range after a flash. Since there is no evidence of a separate mechanism directly inactivating the PDE, it is assumed that all of the active species of the enzyme cascade (Fig. 2), namely $G_{\alpha}$-GTP and R*, must be inactivated to allow PDE deactivation.

$G_{\alpha}$-GTP is inactivated by its intrinsic GTPase activity; after GTP hydrolysis, $G_{\alpha}$-GDP reassociates with $G_{\beta\gamma}$ on the membrane. In order to prevent R* from further binding $G_{\alpha\beta\gamma}$-GDP and catalyzing GDP/GTP exchange, R* must also be deactivated.

The spontaneous decay of metarhodopsin II occurs in the range of minutes; accordingly, the PDE in isolated ROS suspensions remains active for minutes after flash stimulation. It has been shown, however, that deactivation of PDE can be significantly accelerated by adding ATP, and it has been proposed that ATP exerts its quenching effect via phosphorylation of R* (Liebman and Pugh 1980). According to this hypothesis, phosphorylated R* (P-R*) should no longer be capable of activating GDP/GTP exchange on G-proteins, even though it is spectrally still in the metarhodopsin II state. It is not easy to prove or disprove this hypothesis in the rather complex system of ROS membranes since they not only contain a variety of protein kinases besides rhodopsin kinase (Lee et al. 1981, Kapoor and Chader 1984), but also several proteins besides rhodopsin which become phosphorylated in the presence of ATP (Hermolin et al. 1982).

In order to exclude other reactions of ATP, we recently studied the effects of rhodopsin phosphorylation on PDE activity in the *absence* of ATP, by using previously phosphorylated rhodopsin. Rod outer segments were illuminated in the presence of $|\gamma\text{-}^{32}P|$ATP under conditions favoring maximum phosphate incorporation. The chromophore was then "regenerated" by adding excess 11-cis retinal, followed by extensive washing to remove ATP, soluble and peripheral proteins. The resulting washed membranes (P-disks) contained fully regenerated, dark-adapted rhodopsin with 5 - 6 phosphate groups bound per average rhodopsin (Kühn et al. 1984). Purified G-protein and purified PDE were reassociated to P-disks at moderate ionic strength, and the PDE activity induced by a flash of light was monitored using the pH changes associated with cGMP hydrolysis (Yee and Liebman 1978).

Figure 4B shows that flash-illuminated, phosphorylated rhodopsin is still able to activate the PDE (upper curve) but that addition of purified 48 kd-protein almost completely blocks this activating capacity (lower curves). Systematic comparisons between phosphorylated and unphosphorylated disk membranes (not shown in Fig. 4) have shown that phosphorylation by itself already reduces rhodopsin's activating capacity by a factor of 2 - 4; this quenching effect of phosphorylation is potentiated by addition of 48 kd-protein. On the other hand, 48 kd-protein added to unphosphorylated disks has no effect (Wilden et al., in preparation). Both R* phosphorylation and 48 kd-protein are thus required to effectively quench PDE activation.

Similar effects of 48 kd-protein are shown in Figure 4A for the more complex system of ROS membranes which had not previously been phosphorylated. A flash bleaching $2 \times 10^{-4}$ fraction of rhodopsin induced

**Effect of 48 kd-protein on PDE activity**

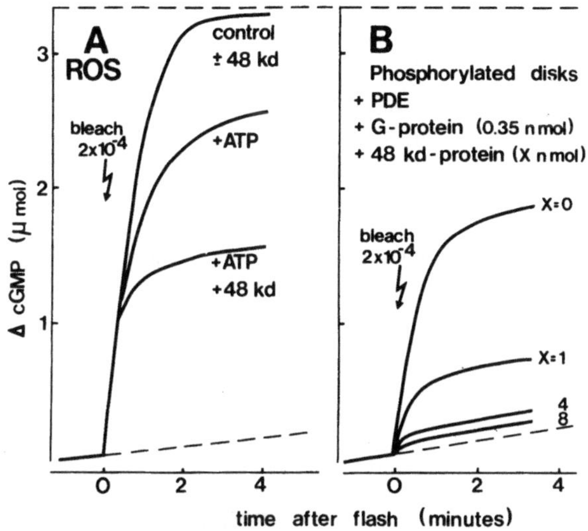


Fig. 4A,B. Phosphodiesterase activity, measured by $H^+$ release accompanying cGMP hydrolysis (Yee and Liebman 1978). Samples contained 3.5 μM rhodopsin, 100 μM GTP, and 2 mM cGMP in 1.5 ml. The reaction was started by bleaching $2 \times 10^{-4}$ fraction of the rhodopsin. *Dashed lines* show basal (dark) PDE activity. A Suspensions of "complete" ROS. Upper trace, PDE activity in the absence of ATP with and without added 48 kd-protein; middle, addition of 500 μM ATP; lower trace, 500 μM ATP plus 1 nmol 48 kd-protein. B Purified enzymes reassociated to previously phosphorylated and regenerated disk membranes; PDE assays in absence of ATP. Constant amounts of G-protein and PDE, and various amounts of 48 kd-protein added as indicated. The quenching effect of phosphorylation on rhodopsin's PDE-activating capacity is potentiated by 48 kd-protein addition

saturating activation of PDE which led to complete hydrolysis of the cGMP pool within 2 min, without any sign of PDE deactivation in the absence of ATP (upper curve); addition of 48 kd-protein had no effect in this case (upper curve). Addition of ATP to ROS significantly accelerated PDE deactivation (middle curve), and this quenching effect of ATP was potentiated by addition of 48 kd-protein (lower curve in Fig. 4A).

What is the mechanism by which 48 kd-protein quenches PDE activation? The most plausible model would be that 48-kd protein binds to P-R* and thereby sterically blocks the access of G-protein to P-R*. It has indeed been shown earlier that although 48 kd-protein binds to illuminated ROS membranes in the absence of ATP (Kühn 1978), it binds much more strongly in the presence of ATP (Kühn 1980), suggesting that it may bind preferentially to phosphorylated R*.

Recent experiments, using the same P-disk preparations as in Figure 4, have in fact shown that 48 dk-protein binds much more strongly to phosphorylated than to unphosphorylated disks, if only 5 - 10% of the rhodopsin is bleached (Kühn et al. 1984). Moreover, the 48 kd-protein bound to partially bleached P-disks can be displaced by adding purified G-protein, indicating competition between G-protein and 48 kd-

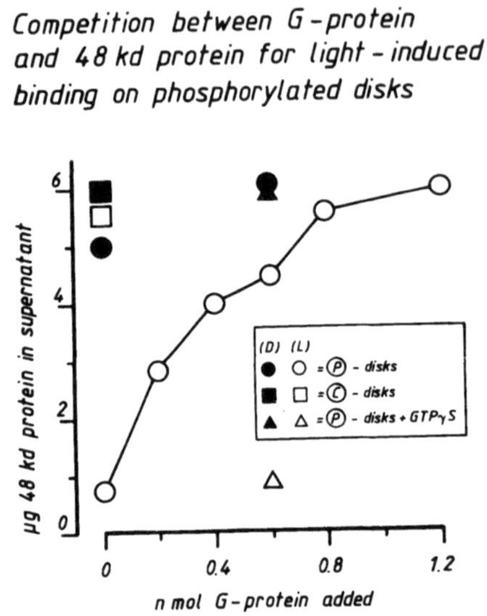


Fig. 5. Light-induced binding of 48 kd-protein to phosphorylated disk membranes (*"P-disks"*), and its displacement by added purified G-protein. Samples contained constant amounts of rhodopsin (350 μg each) and of 48 kd-protein (7 μg each) but various amounts of G-protein as indicated on the abscissa. Suspensions were either kept dark (*filled symbols*) or shortly illuminated (5% rhodopsin bleached) before centrifugation. The amount of 48 kd-protein in the supernatants was determined by gel densitometry and is plotted on the ordinate. Unphosphorylated control disks (*"C-disks"*) do not show light-induced binding of 48 kd-protein under these conditions. (Kühn et al. 1984)

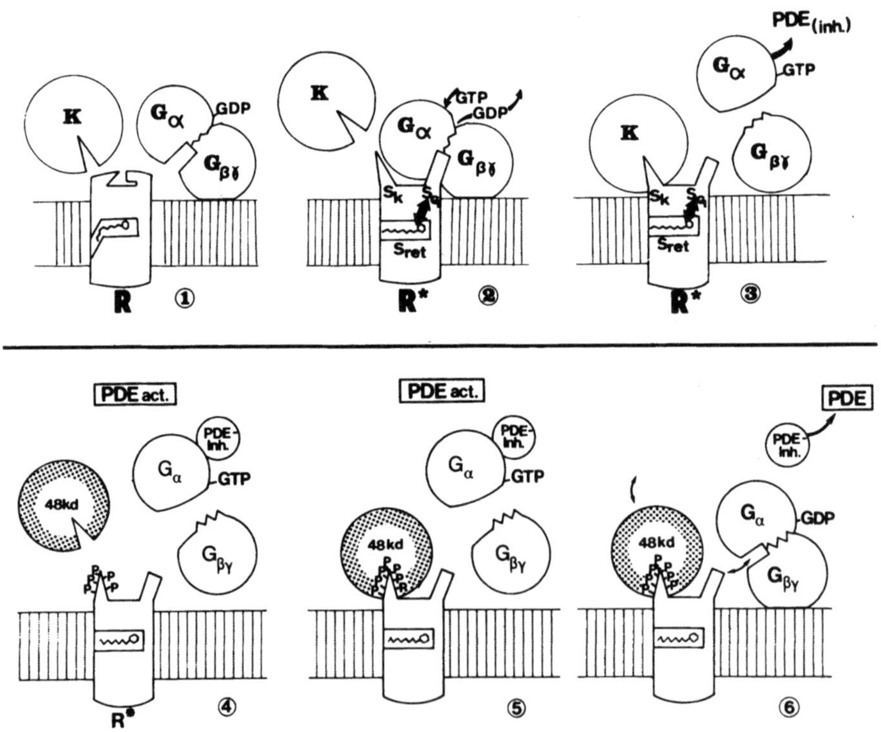


Fig. 6. Scheme of protein-protein interactions at six consecutive stages (numbered 1-6), leading first to light-induced activation and then to deactivation of the PDE enzyme cascade. *R* dark-adapted rhodopsin; *R** photoexcited rhodopsin; *K* rhodopsin kinase; $G_\alpha$, $G_{\beta\gamma}$ G-protein. For details see text

protein for binding to P-R* (Fig. 5). It therefore seems likely that, vice versa, an excess of 48 kd-protein prevents G-protein from interacting with P-R*. In intact rod cells, the concentration of 48 kd-protein is nearly as high as that of G-protein. A 500 - 1000 fold dilution of the system, necessary for the experiments of Figures 4 and 5, diluted only the soluble 48 kd-protein but not the membrane-associated G-protein with respect to rhodopsin. This differential dilution probably explains why such a large excess of 48 kd-protein is required to effectively quench the PDE (see Fig. 4B); in the living cell, the situation is much more in favor of the 48 kd-protein.

Figure 6 schematically summarizes the protein-protein interactions that lead first to activation and later to deactivation of PDE. Photoexcited rhodopsin (R*) exhibits two sites of interaction, one for G-protein and one for rhodopsin kinase. Binding of G-protein to R* (state 2 in Fig. 6) enables the exchange of GTP for GDP on the $G_\alpha$ subunit, followed by dissociation of $G_\alpha$-GTP from $G_{\beta\gamma}$ and from R* (state 3). $G_\alpha$-GTP activates the PDE by binding an inhibitory subunit of the PDE (state 4). One R* can sequentially interact with hundreds of G-protein molecules and catalyze their GDP/GTP exchange. The complex R*-$G_{\alpha\beta\gamma}$-GDP (state 2) can be artificially stabilized by omitting GTP from the buffer; under this condition, access of the kinase to R* is sterically blocked (Kühn 1984). Normally, i.e., in the presence of GTP, the lifetime of the complex R*-$G_{\alpha\beta\gamma}$ is less than a millisecond (Vuong et al. 1984); during the rapid cycles of G-protein binding and dissociation, R* is "free" for short periods of time to bind the kinase which then incorporates phosphate groups into R* (states 3 and 4). Since the rod cell contains much more G-protein than rhodopsin-kinase (about 100 times more), phosphorylation of R* will lag somewhat behind G-protein activation. Phosphorylated R* then is less capable – and when it binds 48 kd-protein even much less capable – of activating G-protein (stages 5 and 6). The deactivation of R* by phosphorylation and 48 kd-protein binding is paralleled by the deactivation of $G_\alpha$-GTP via GTP hydrolysis and reassociation of $G_\alpha$-GDP to $G_{\beta\gamma}$ (stage 6). The original dark-adapted state is restored only after subsequent dephosphorylation of opsin, and after regeneration of its chromophore 11-cis retinal (not shown). This scheme is, of course, highly simplified and leaves many open questions for future research.

*Acknowledgments.* We thank E. Wüst and R. Esser for technical assistance, A. Einerhand for drawing figures, and A. Eckert for typing the manuscript. Part of the work was supported by a grant from SFB 160, Deutsche Forschungsgemeinschaft. One of us (S.W.H.) received a DAAD fellowship.

## References

Baehr W, Devlin MJ, Applebury ML (1979) Isolation and characterization of cGMP phosphodiesterase from bovine rod outer segments. J Biol Chem 254:11669-11677

Bownds D, Dawes J, Miller J, Stahlman M (1972) Phosphorylation of frog photoreceptor membranes induced by light. Nature (London) 237:125-127

Chen YA, Hubbell WL (1978) Reactions of the sulfhydryl groups of membrane-bound bovine rhodopsin. Membr Biochem 1:107-130

Dorey C, Cozette J, Faure JP (1982) A simple and rapid method for isolation of retinal S-antigen. Opthalmic Res 14:249-255

Emeis D, Kühn H, Reichert J, Hofmann KP (1982) Complex formation between metarhodopsin II and GTP-binding protein in bovine photoreceptor membranes leads to a shift of the photoproduct equilibrium. FEBS Lett 143:29-34

Fesenko EE, Kolesnikov SS, Lyubarsky AL (1985) Induction by cyclic GMP of cationic conductance in plasma membrane of retinal rod outer segment. Nature (London) 313:310-313

Fung BKK (1983) Characterization of transducin from bovine retinal rod outer segments. Separation and reconstitution of the subunits. J Biol Chem 258:10495-10502

Fung BKK, Stryer L (1980) Photolyzed rhodopsin catalyzes the exchange of GTP for bound GDP in retinal rod outer segments. Proc Natl Acad Sci USA 77:2500-2504

Fung BKK, Hurley JB, Stryer L (1981) Flow of information in the light-triggered cyclic nucleotide cascade of vision. Proc Natl Acad Sci USA 78:152-156

Hermolin J, Karell MA, Hamm HE, Bownds MD (1982) Calcium and cyclic GMP regulation of light-sensitive protein phosphorylation in frog photoreceptor membranes. J Gen Physiol 79:633-655

Hofmann KP, Schleicher A, Emeis D, Reichert J (1981) Light-induced axial and radial shrinkage effects and changes of the refractive index in isolated bovine rod outer segments and disk vesicles. Biophys Struct Mech 8:67-93

Hurley JB, Stryer L (1982) Purification and characterization of the regulatory subunit of the cyclic GMP phosphodiesterase from retinal rod outer segments. J Biol Chem 257:11094-11099

Kapoor CL, Chader GJ (1984) Endogenous phosphorylation of retinal photoreceptor outer segment proteins by calcium phospholipid-dependent protein kinase. Biochem Biophys Res Commun 122:1397-1403

Kühn H (1978) Light-regulated binding of rhodopsin kinase and other proteins to cattle photoreceptor membranes. Biochemistry 17:4389-4395

Kühn H (1980) Light- and GTP-regulated interaction of GTPase and other proteins with bovine photoreceptor membranes. Nature (London) 283:587-589

Kühn H (1981) Interactions of rod cell proteins with the disk membrane: Influence of light, ionic strength, and nucleotides. Curr Top Membr Transp 15:171-201

Kühn H (1982) Light regulated binding of proteins to photoreceptor membranes, and its use for the purification of several rod cell proteins. Methods Enzymol 81: 556-564

Kühn H (1984) Interactions between photoexcited rhodopsin and light-activated enzymes in rods. Prog Retinal Res 3:123-156

Kühn H, Dreyer WJ (1972) Light-dependent phosphorylation of rhodopsin by ATP. FEBS Lett 20:1-6

Kühn H, Hargrave PA (1981) Light-induced binding of GTPase to bovine photoreceptor membranes: Effects of limited proteolysis of the membranes. Biochemistry 20: 2410-2417

Kühn H, Bennett N, Michel-Villaz M, Chabre M (1981) Interactions between photoexcited rhodopsin and GTP-binding protein: kinetic and stoichiometric analysis from light-scattering changes. Proc Natl Acad Sci USA 18:6873-6877

Kühn H, Mommertz O, Hargrave PA (1982) Light-dependent conformational change at rhodopsin's cytoplasmic surface detected by increased susceptibility to proteolysis. Biochim Biophys Acta 679:95-100

Kühn H, Hall SW, Wilden U (1984) Light-induced binding of 48 kDa protein to photoreceptor membranes is enhanced by phosphorylation of rhodopsin. FEBS Lett 176: 473-478

Lee RH, Brown BM, Lolley RN (1981) Protein kinases of retinal rod outer segments: identification and partial characterization of cyclic nucleotide dependent protein kinase and rhodopsin kinase. Biochemistry 20:7532-7538

Liebman PA, Pugh EN (1980) ATP mediates rapid reversal of cyclic GMP phosphodiesterase activation in visual receptor membranes. Nature (London) 287:734-736

Miller WH (ed) (1981) Molecular mechanisms of photoreceptor transduction. Curr Top Membr Transp 15

Pellicone C, Nullans G, Virmaux N (1985) Localization of light-induced conformational changes in bovine rhodopsin. FEBS Lett 181:179-183

Pfister C, Kühn H, Chabre M (1983) Complex formation between photoexcited rhodopsin and GTP-binding protein influences the post metarhodopsin II decay and the phosphorylation rate of rhodopsin in frog rods. Eur J Biochem 136:489-499

Pfister C, Chabre M, Plouet J, Tuyen VV, DeKozak Y, Faure JP, Kühn H (1985) Retinal S antigen identified as the 48 K protein regulating light-dependent phospodiesterase in rods. Science 228:891-893

Uchida S, Wheeler GL, Yamazaki A, Bitensky MW (1981) A GTP-protein activator of phosphodiesterase which forms in response to bleached rhodopsin. J Cycl Nucleotide Res 7:95-104

Vuong TM, Chabre M, Stryer L (1984) Millisecond activation of transducin in the cyclic nucleotide cascade of vision. Nature (London) 311:659-661

Wald G (1968) Molecular basis of visual excitation. Science 162:230-239

Wheeler GL, Bitensky MW (1977) A light-activated GTPase in vertebrate photoreceptors: Regulation of light-activated cyclic-GMP phosphodiesterase. Proc Natl Acad Sci USA 74:4238-4242

Wilden U, Kühn H (1982) Light-dependent phosphorylation of rhodopsin: Number of phosphorylation sites. Biochemistry 21:3014-3022

Yee R, Liebman PA (1978) Light-activated phosphodiesterase of the rod outer segment. Kinetics and parameters of activation and deactivation. J Biol Chem 253:8902-8909

# The Molecular Biology of Muscle and Brain Acetylcholine Receptors

J. F. Jackson[1], D. M. W. Beeson[1,2], B. M. Conti-Tronconi[3], V. B. Cockcroft[1], T. L. Anderton[1], L. D. Bell[2], A. F. Wilderspin[1], and E. A. Barnard[1]

## Introduction

Nicotinic acetylcholine receptors (AChR) mediate chemical communication at synapses in many parts of the vertebrate nervous system, including neuromuscular junctions, autonomic ganglia, and certain sites in the brain. This occurs through the interaction of neuronally released acetylcholine (ACh) with recognition sites on the AChR in the post-synaptic membrane. Binding of ACh to the AChR activates a gated cation channel and results in a transient change in the permeability of the membrane, which can be measured as a depolarisation of the transmembrane electrical potential. This has been directly demonstrated for the best characterised AChR, that from the electroplax of *Torpedo* sp. or *Electrophorus* sp.

Advances in the molecular biology of vertebrate AChR have depended greatly on (1) the use of the electroplax as a rich source of pure receptor protein (100 mg/kg) and (2) availability of snake polypeptide neurotoxins, particularly alpha-bungarotoxin (α-BuTX) as selective ligands that bind pseudo-irreversibly to the AChR. Therefore, we will first discuss the contribution of molecular biology to our understanding of *Torpedo* AChR structure, evolution and function.

## The *Torpedo* AChR: A Multisubunit Transmembrane Protein

Affinity chromatography using α-BuTX coupled to Sepharose can be used to prepare receptor from solubilised membranes of the *Torpedo* electric organ. The AChR consists of four different subunits – alpha ($M_r$ 40,000), beta (50,000), gamma (60,000), and delta (65,000) – present in the stoichiometry $\alpha_2\beta\gamma\delta$, a value consistent with the observed molecular weight for native receptor of 275,000 (Fig. 1). Affinity labelling experiments demonstrate that an ACh binding site is present on each α-subunit (Damle and Karlin 1978, Wolosin et al. 1980). X-ray diffraction studies and experiments using proteases and receptor-specific antibodies have shown that the receptor is a membrane-spanning protein with approximately 50% of the mass projecting above the lipid bilayer. The ion channel is thought to have a cylindrical cross-section with the extracellular protein (containing the AChR binding sites) seen as a funnel. This is most dramatically visualized in a negative stain

[1]Department of Biochemistry, Imperial College of Science and Technology, London SW7 2AZ, UK

[2]Searle Research and Development, PO Box 53, Lane End Road, High Wycombe, Buckinghamshire, HP12 4HL, UK

[3]Division of Chemistry, California Institute of Technology, Pasadena, CA 91125, USA

36. Colloquium - Mosbach 1985
Neurobiochemistry

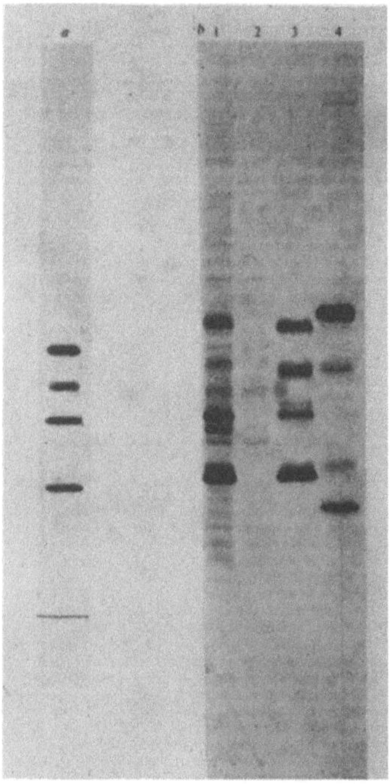

Fig. 1. Subunit composition of the *Torpedo marmorata* AChR as obtained from the purified protein or by translation of mRNA in vivo. a Photograph of a stained SDS gel of *Torpedo* AChR subunits (note that this gel is shorter than that in b). b Autoradiograph of an SDS-polyacrylamide gel: *Lane 1* AChR subunits purified by α-BuTX-Sepharose chromatography from the membranes of *Xenopus* oocytes microinjected with *Torpedo* electric organ mRNA and labelled with [$^{35}$S]methionine; *lane 2* as lane 1 but with preincubation of solubilised membranes with excess α-BuTX before chromatography; *lane 3* [$^{3}$H]propionylated *Torpedo* AChR prepared from solubilised electric organ membranes; *lane 4* iodinated molecular weight markers. (Sumikawa et al. 1981)

micrograph from the laboratory of R. Stroud taken at the edge of a folded-over membrane vesicle (Fig. 2). X-ray scattering data from the same laboratory demonstrates that (1) the protein extends 55A extracellularly and 15A intracytoplasmically and (2) has a large α-helical content oriented perpendicularly to the plane of the membrane. Electron microscopy of uranyl acetate-stained membrane sheets dispersed by base treatment and sonication reveals pairs of stained rosettes which represent dimeric AChR molecules known to be joined together by disulphide bonds through the delta subunit (Fig. 3). From the stain density it was concluded that the uranyl acetate penetrates through the entire transmembrane length of the receptor at a single ion channel.

## Cloning the *Torpedo* AChR Genes

How have the genes encoding each of the subunits of *Torpedo* AChR been isolated? Cloning of the AChR represented a difficult problem because the protein consisted of four different polypeptides encoded by separate mRNA's (D.J. Anderson and Blobel 1981). Gene isolation depended on the relatively high abundance of mRNA for the subunits (each approximately 0.5% of electroplax mRNA; Mendez et al. 1980), amino terminal protein sequence data (Raftery et al. 1980), subunit-specific antibodies to each of the four chains (Claudio and Raftery 1977, Lindstrom et al. 1978), and the ability of the *Xenopus* oocyte to faithfully translate, process and assemble functional AChR when injected with electroplax mRNA (Fig. 1). Two distinct approaches were independently employed by at least four groups to obtain the genes. Both strategies aimed to isolate the clones from a complementary DNA (cDNA) library representing all molecules in the population of electroplax mRNA. In such a library, AChR clones would be present at a frequency similar to their

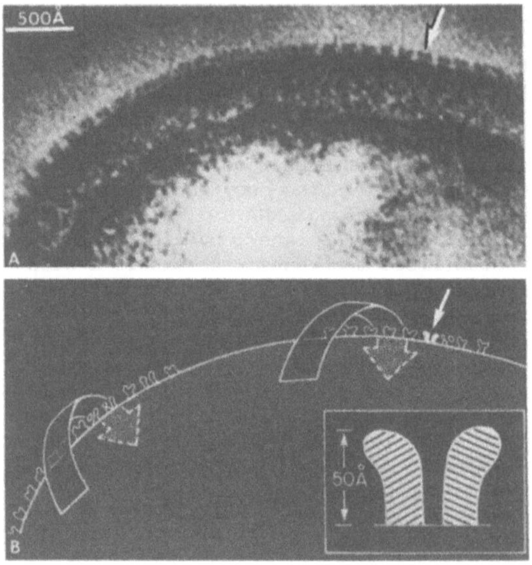


Fig. 2 A,B. View of a uranyl acetate-stained synaptic edge of a *Torpedo* electric organ membrane vesicle (A) and a schematic representation of the 55A funnel-shaped protrusions of the receptor molecules (B). (Kistler et al. 1982)

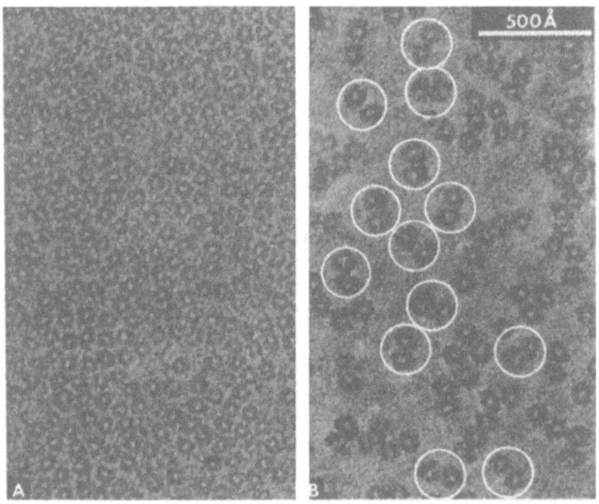


Fig. 3 A,B. Electron micrograph of uranyl acetate-stained membrane sheets from *Torpedo* electric organ. The closely packed AChR molecules (A) have been dispersed by base treatment and sonication (B). *Circles* outline some of the dimeric AChR molecules. (Kistler et al. 1982)

abundance in the mRNA population (0.5% or 1 in 200) rather than at a frequency reflecting their representation in the genome (1 in 1,000,000) as they would be in a library prepared from chromosomal DNA (a genomic library).

We were able to show that if total electroplax mRNA were sedimented on a sucrose gradient, that a fraction 2000 bases (2 kb) in size was active in the oocyte assay, and must therefore contain mRNA for all four subunits. This led to use of a preliminary gradient step to generate a fraction enriched in AChR mRNA. Protein sequence information corresponding to the amino terminus of the mature alpha subunit was used to construct synthetic oligonucleotide probes corresponding to all possible triplets coding for a four-amino-acid-long segment (Fig. 4). These oligonucleotides were end-labelled with [$^{32}$P]ATP and used to screen the cDNA library; positive hybridising clones were

(a) **Amino acid sequence** .....Val.Glu.His.His (residues 22 23 24 25)

(b) **Possible codons** 5'GU[A/C/U/G]GA[A/G].CA[U/C].CA[U/C] 3'

(c) **Primers**

| | |
|---|---|
| α1 | 3'CAT.CTT.GTG.GTG 5' |
| α2 | 3'CAT.CTC.GTG.GTG 5' |
| α3 | 3'CAG.CTT.GTG.GTG 5' |
| α4 | 3'CAG.CTC.GTG.GTG 5' |
| α5 | 3'CAA.CTT.GTG.GTG 5' |
| α6 | 3'CAA.CTC.GTG.GTG 5' |

Fig. 4. Deduction of synthetic oligonucleotide probes for amino acids 22-25 of the α-subunit of *Torpedo* AChR (a). c represents a subset of all possible nucleotide sequences able to code for these residues and was shown to contain the actual coding sequence by a primer extension experiment. (Sumikawa et al. 1982)

isolated, sequenced and the sequence compared to the α-subunit protein sequence. This cDNA clone contained a small DNA insert which coded for part of the amino terminus of the α-subunit, and was used as a hybridisation probe to isolate longer clones from the library which contained the entire coding sequence (Sumikawa et al. 1982). An essentially identical approach was used by the laboratory of S. Numa to isolate full-length cDNA clones for all four *Torpedo* AChR subunits (Noda et al. 1982, 1983a).

An alternative cloning strategy isolates genes expressed in electroplax, but not in brain, by hybridising replicate filters from the cDNA library to either electroplax or brain total [$^{32}$P] cDNA and selecting clones which hybridise with the former but not the latter. Clones identified by this "plus-minus" screen were tested for their ability to select, by DNA-RNA hybridisation, AChR subunit mRNA. Messenger RNA's hybridising to a clone were eluted, translated in vitro, and the protein products analysed by immunoprecipitation with anti-AChR subunit antisera and subsequent gel electrophoresis. This method resulted in the isolation of alpha-subunit (Devillers-Thiéry et al. 1983) and gamma-subunit (Ballivet et al. 1982) clones.

## Structure, Orientation, and Evolution of the Four AChR Subunits

When the deduced amino acid sequences of the four *Torpedo* AChR subunits are compared, the amino terminal sequence homology previously observed by Raftery et al. (1980) is seen to extend, to a greater or lesser extent, through the entire polypeptide chain. In addition. the alpha/beta and gamma/delta gene pairs show higher homology, supporting the contention that the genes for the four subunits have evolved by duplication and divergence from a single primitive gene. Presumably the primitive receptor consisted of a homopolymer of identical subunits, like the gap junction (Unwin and Zampighi 1980). The contemporary vertebrate receptor, containing four types of chain, is thought to have evolved in two separate gene duplication events, each followed by divergence. Initially, the primitive gene duplicated, then alpha/beta-like and gamma/delta-like genes evolved. After a second duplication, alpha diverged from beta and gamma diverged from delta.

The deduced amino acid sequences of the four subunits show them to contain alternating stretches of hydrophilic and hydrophobic amino acids. Standard methods of empirical protein secondary structure prediction and hydrophilicity analysis has led to models for a pseudosymmetric orientation of all five subunits in the cell membrane. There is some disagreement as to whether each polypeptide crosses the membrane four or five times. We prefer a model proposed by Guy (1984) and Finer-Moore and Stroud (1984) in which the membrane is traversed a total of five times – by four highly conserved hydrophobic alpha-helices and one less highly conserved amphipathic alpha-helix as shown (Fig. 5): an amphipathic helix has hydrophobic residues along one face and hydrophilic groups along another face of the helix, as viewed from above. The former are thought to comprise part of the insulation and the latter a portion of the lining of the ion channel. The insulation and lining would be comprised of equivalent regions from each of the five receptor subunits. This model, in which the large amino terminal domain of a subunit lies to the outside and the carboxy terminus on the inside of the postsynaptic membrane is consistent with a variety of biophysical, biochemical, and immunological data. Data from two laboratories (Lindstrom et al. 1984, Young et al. 1985) have shown that antibodies raised against carboxy-terminal synthetic peptides can bind to receptor in *Torpedo* membrane vesicles only if they are first treated with saponin, an agent which forms holes in the vesicle membrane. The prototypical site-directed mutagenesis studies of the Numa laboratory also support this notion. They show that deletion of the putative amphipathic transmembrane region in a cloned alpha AChR gene and incorporation of this mutant polypeptide into an otherwise normal receptor (i.e., containing wild-type beta, gamma, and delta chains) abolishes channel function (Mishina et al. 1985).

## The Vertebrate Muscle AChR: A Homologous Protein

Though present at a concentration three orders of magnitude lower than that in the electric organ, the vertebrate muscle AChR is very similar in structure and function. AChR purified from denervated chicken pectoral muscle (where the AChR concentration is 20- to 40-fold higher than in innervated muscle), when complexed with [$^{125}$I]α-BuTX, cosediments with the 9S complex from *Torpedo*. SDS gel electrophoresis of this AChR emphasises the overall similarity to *Torpedo* AChR: a 43K alpha subunit which binds toxin and can be affinity labelled with bromoacetylcholine (Br-ACh); a 50K beta subunit, a 55K gamma subunit, and a 57K delta subunit are present in the usual stoichiometry (Fig. 6). Aminoterminal

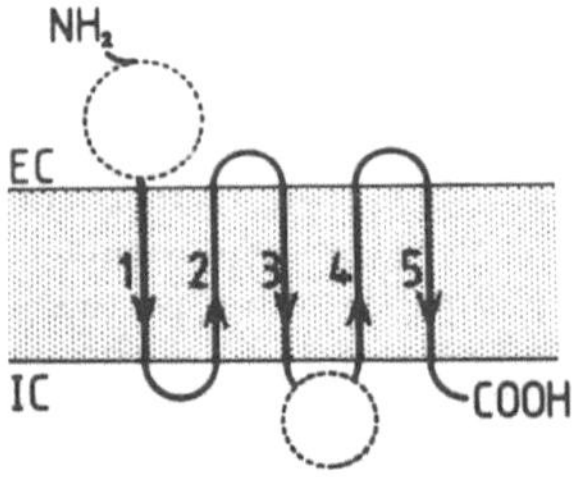


Fig. 5. A model for the arrangement of one of the AChR subunits in the synaptic membrane, based on interpretation of the secondary structure of the α-subunit of *Torpedo* AChR as predicted from the deduced amino acid sequence. In this orientation, there is a large amino terminal non-helical segment located on the extracellular face (*EC*) and the carboxy terminus is located on the intracellular face (*IC*). (Dolly and Barnard 1984)

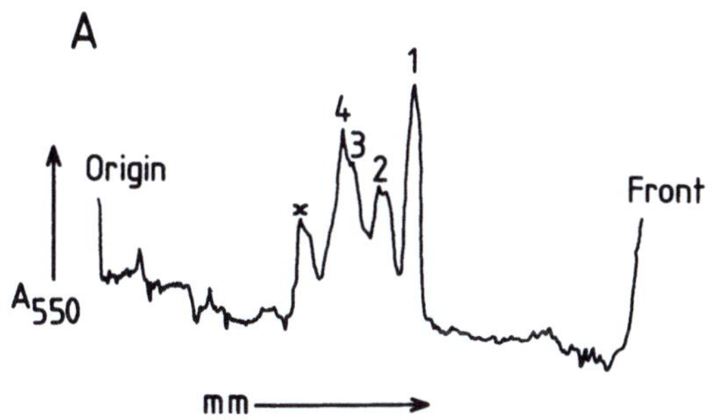


Fig. 6. Subunit composition of chicken muscle AChR. Polypeptides present in an eluate of the toxin-Sepharose column were electrophoresed on an SDS gel. Gels stained with Coomassie Brilliant Blue and scanned at 550 nm gave the pattern illustrated. The band marked *x* is an artefact resulting from SDS elution, as it is not seen when AChR is eluted biospecifically with carbamylcholine. Partial separation of peaks *3* and *4* is reproducibly observed. (Beeson et al. 1985)

**α-Subunit**

```
                    10                  20
  1 2 3 4 5 6 7 8 9 0 1 2 3 4 5 6 7 8 9 0 1 2 3 4 5
  -------------------------------------------------
C X E H E T R L V D D L F R D Y S K V V R P V E N H
  -------------------------------------------------

H S E H E T R L V A K L F K D Y S S V V R P V E D H
B S E H E T R L V A K L F E D Y N S V V R P V E D H
T S E H E T R L V A N L L E N Y N K V I R P V E H H
E S E D E T R L V K N L F S G Y N K V V R P V N H F

 26      30                  40
  6 7 8 9 0 1 2 3 4 5 6 7 8 9 0 1 2 3 4 5
  ---------------------------------------
C R D A V K V X V G L A L I X L I N V D E
  ---------------------------------------

H R Q V V E V T V G L Q L I Q L I N V D E
B R Q A V E V T V G L Q L I Q L I N V D E
T T H F V D I T V G L Q L I Q L I S V D E
E K D A V V V T V G L
```

Fig. 7. Amino terminal acid sequence of α AChR subunits. The chemically determined amino acid sequence of the chicken muscle α-subunit (*C*) is compared to those of the human (*H*), *Torpedo* (*T*), *Electrophorus* (*E*), and bovine (*B*) cognates. Residues present in all species are underlined. These and similar data for the other three subunits are presented in Beeson et al. (1985)

microsequence analysis demonstrates that each subunit is strongly homologous to the *Torpedo* counterpart (Fig. 7; Beeson et al. 1985), and three of the four sequences have, as we shall see, been confirmed by cloning (Nef et al. 1984, Beeson et al. 1985).

## Chicken AChR Gene Cloning

The strong sequence conservation between vertebrate and fish AChR subunits suggested a simple cloning strategy, in which *Torpedo* probes are hybridised to vertebrate cDNA or genomic DNA libraries under conditions of reduced stringency, where a certain amount of base pair mismatching

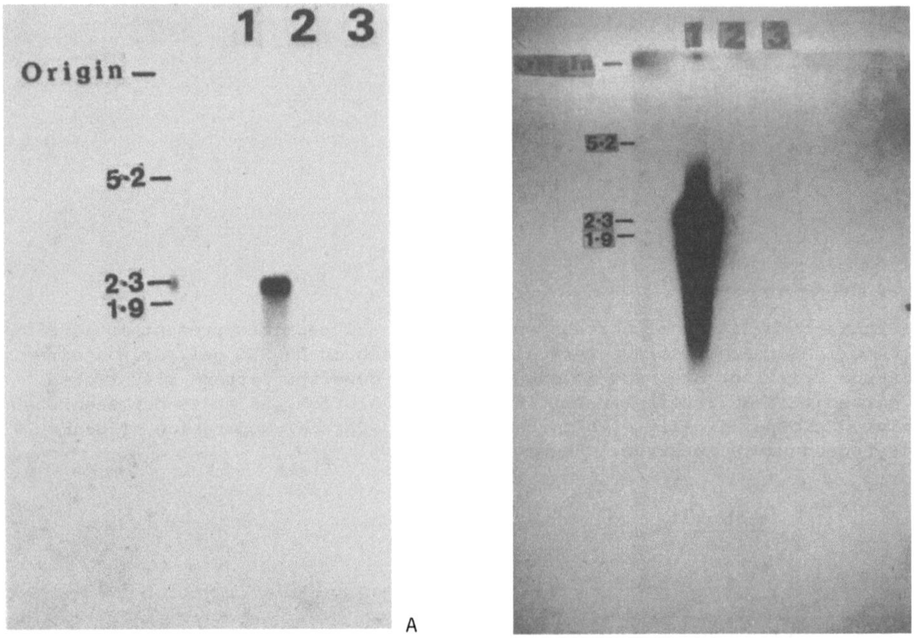


A B

Fig. 8 A,B. Detection of chicken AChR mRNA by cross-hybridisation with a *Torpedo* probe. Poly(A)$^+$ RNA's from *Torpedo* electric organ (*lane 1* 1 μg), embryonic chicken muscle (*lane 2* 20 μg) and Marek's virus transformed chicken T lymphocytes (*lane 3* 20 μg) were electrophoresed on a formaldehyde-agarose gel, transferred to nitrocellulose and hybridised to a nick translated *Torpedo marmorata* cDNA clone corresponding to the transmembrane regions M1 and M2 of the α-subunit. A and B are different exposures of the same autoradiograph

is tolerated. A *Torpedo* α-subunit cDNA clone can detect related sequences in embryonic chicken muscle mRNA but not in chicken T lymphocyte mRNA (Fig. 8). In fact, the low hybridisation signal seen when chicken muscle is compared to *Torpedo* electroplax mRNA is underemphasised by the loading of twenty times less mRNA in the latter lane. Ballivet et al. (1983) also used *Torpedo* subclones corresponding to what was thought to be the most highly conserved region of the polypeptide – DNA encoding the transmembrane regions M1 and M2 (which turn out to have, with the exception of one conservative amino acid substitution, *identical* primary sequences when *Torpedo*, calf, and human are compared) to isolate the corresponding exons of the chicken muscle alpha subunit from a genomic library. This group recently reported isolation of a chicken genomic clone containing linked, closely apposed delta and gamma structural genes (Nef et al. 1984). Noda et al. (1983b) used a similar procedure to isolate calf cDNA and human genomic DNA clones for the same subunit. In our hands, this strategy, in conjunction with rescreening using oligonucleotides derived from the amino terminus of the chicken α-subunit, yielded full-length cDNA clones from an embryonic pectoral muscle cDNA library (Beeson et al. 1985).

The 1812 nucleotide long sequence obtained contains a single long open reading frame which was used to deduce the complete amino acid sequence of the α-subunit. After a segment corresponding to the signal sequence, the deduced amino acid sequence was found to match the chemically determined N-terminal amino acid sequence. Only one difference (glu in the gene at position 14 instead of the asp predicted) is found between these two in the first 29 residues. The mature subunit has 437 residues, as do all three other fully sequenced α-subunit genes (Noda et al. 1983b) and a deduced molecular weight of 50,095. A comparison of the N-terminal amino acid sequences of chicken and *Torpedo* alpha AChR (a relatively non-conserved portion of the polypeptide) clearly illustrates the strong conservation between vertebrate and fish (Fig. 7). We have also isolated, but not completely sequenced, cDNA clones corresponding to the other three subunits of this AChR.

## Receptor Biology

These DNA probes are being used in two studies which are currently in progress in our laboratories: (1) regulation of receptor biosynthesis during embryonic development and following denervation and (2) isolation of central nervous system AChR genes. Northern blots show that mature α-subunit mRNA from embryonic, denervated or normal (innervated) muscle migrates as a single species of 3.1 kb. Upon longer exposure of the autoradiogram additional bands of 4.2 kb and 5.9 kb can be seen in all three lanes (unpublished observations). These larger species probably represent incompletely processed primary transcripts (nuclear RNA molecules containing unexcised intervening sequences) and have been noted by others (Takai et al. 1984, Ballivet, personal communication). In any case, integration of densitometric scans of the autoradiogram reveal the three size classes are present in identical proportions in each of the three samples.

The extrajunctional AChR that appears upon denervation is known to be newly synthesised and not recruited from either a reserve pool of inactive receptor or by redistribution of synaptic receptor. Regulation of this new net synthesis could occur at one or more of many points. Insertion of AChR in the muscle plasma membrane is the last step in a long synthetic pathway which includes transcription of the four structural genes, processing and polyadenylation of nuclear RNA, transport of mRNA to the cytoplasm, translation of mRNA, co-translational modification within the rough endoplasmic reticulum, posttranslational modification within the Golgi apparatus and assembly of AChR subunits. Either by densitometric scans of Northern blots or "dot-blots", we find that denervation of chicken pectoral muscle results in a 20-fold increase in the steady-state level of α-subunit mRNA, as compared to normal pectoral muscle. A similar increase is seen upon denervation of chicken leg muscle (unpublished observations). Messenger RNA yield per gram of tissue and mRNA purity, as assayed by hybridisation to tritiated poly(U), is similar in all cases. Merlie et al. (1984), using Northern blot analysis, have reported a 100-fold increase in the steady-state level of mouse α-subunit mRNA following denervation of the hind limb. The reason for this discrepancy is not clear, but could simply reflect differences in the physiology of the muscles or organisms used. We are currently extending this analysis to the other three subunits. Since the increase in α-subunit mRNA observed upon denervation parallels that of the receptor protein as assayed by toxin binding, this accumulation is due either to mRNA stabilisation, increased RNA transcription, or both. Run-off assays using isolated nuclei can distinguish between these possibilities (Greenberg and Ziff 1980).

The absolute level of α-subunit sequences in denervated muscle mRNA, 0.017%, was derived by coelectrophoresis of denervated muscle mRNA with precisely measured amounts of pure rabbit globin mRNA (which migrates at 0.7 kb) followed by hybridisation of the Northern blot with a 1:1 mixture of globin and AChR probes of identical specific activity. The abundance of AChR mRNA in innervated muscle can thus be calculated: AChR α-subunit mRNA, like the receptor protein, is present in innervated muscle at a concentration three logs lower than in *Torpedo* electroplax, 0.0009% versus 0.5% (unpublished observations).

## AChR in the Central Nervous System?

Some regions of the vertebrate brain contain an AChR with pharmacological characteristics similar but not identical to those of muscle AChR (Morley and Kemp 1981, Oswald and Freeman 1981). Additionally, some brain areas and peripheral ganglia contain high affinity binding sites for α-BuTX. However, it is by no means generally accepted that the toxin binding components represent functional AChR. In amphibian and avian optic lobe (Oswald and Freeman 1981), human medulloblastoma cells (Syapin et al. 1982) and some sympathetic ganglion sites (Marshall 1981, Jacob and Berg 1983) there is evidence for this identity, even though α-BuTX binding does not always block receptor function. We (Norman et al. 1982) and others (Betz and Pfeiffer 1984) have isolated the α-BuTX binding protein from avian optic lobe, where α-BuTX does block AChR function (Oswald and Freeman 1981). A variety of experiments indicate that the chicken optic lobe α-BuTX binding protein is an AChR structurally homologous to, but *significantly different* from peripheral AChR.

## The Optic Lobe AChR

Toxin binding experiments detect α-BuTX binding in solubilised chicken optic lobe membranes at a concentration comparable to that in innervated skeletal muscle (Wang et al. 1978). This component from optic lobe shows high-affinity binding to iodinated α-BuTX with specific activities of 4000-6000 nmol/g protein. The component purified from "brain" (remainder of the brain after optic lobe removal) showed similar binding characteristics and specific activity. The optic lobe AChR, purified by α-BuTX-Sepharose affinity chromatography, sediments in sucrose gradients as a peak with a sedimentation coefficient of 10.1S – distinctly faster than muscle AChR centrifuged in parallel gradients (Fig. 9). The molecular size of optic lobe AChR is significantly larger than muscle AChR, as shown by hydrodynamic determination with $D_2O$ correction for bound detergent (data not shown).

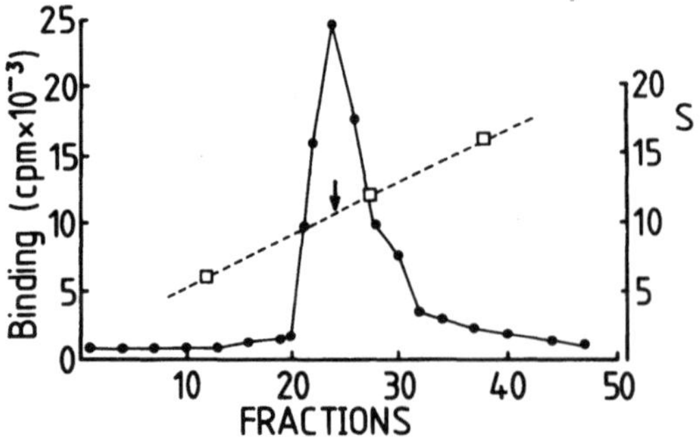


Fig. 9. Sucrose gradient centrifugation of optic lobe AChR. Affinity-purified receptor was centrifuged on a 5%-20% sucrose gradient, fractionated and assayed for binding of $[^{125}I]$α-BuTX. Calibration standards were ADH, catalase and β-galactosidase, with $s_{20,w}$ values on the right hand scale. In four experiments with different preparations the receptor peak was centered (*arrow*) at a mean of 10.1S. (Conti-Tronconi et al. 1985)

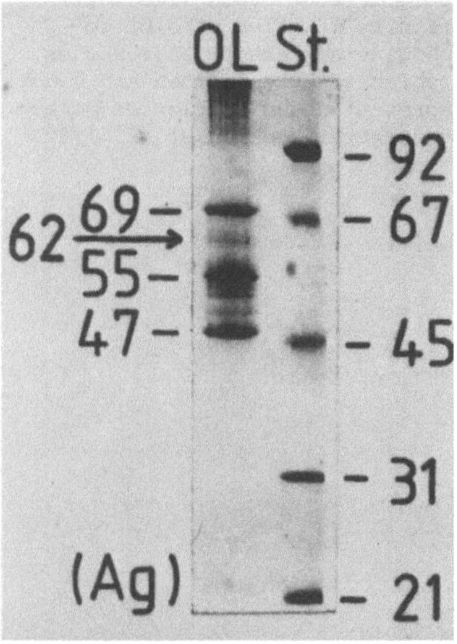


Fig. 10. Subunit composition of optic lobe AChR. Optic lobe AChR (*OL*) was electrophoresed on an SDS gel together with molecular weight markers (*St.*). Proteins were visualised by silver staining. (Conti-Tronconi et al. 1985)

Figure 10 shows a silver stained SDS gel of purified optic lobe AChR. In different experiments five major components consistently appear, of apparent $M_r$ 48K, 56K (doublet), 62K, and 69K. This pattern differs from that previously reported (Norman et al. 1982, Betz and Pfeiffer 1984) and results from the use of an improved protocol for AChR isolation (Conti-Tronconi et al. 1985). What is the evidence that these polypeptides are homologous to peripheral AChR? We had previously shown that the 56K polypeptide could be affinity-labelled using bromoacetylcholine. Affinity labelling of intact receptor from optic lobe or brain specifically visualises the 56K subunit (Fig. 11). Likewise, iodinated α-BuTX can be cross-linked to this subunit using dimethyl suberimidate. Thus, this subunit carries a high-affinity site for AChR and for toxin, as does the α-subunit of the peripheral AChR.

Additional evidence that at least some of the other polypeptides are components of the AChR was obtained by immunoprecipitation. A monoclonal antibody, 7B2, raised against chick muscle AChR, cross-reacts with the α-BuTX binding protein from chicken optic lobe (Mehraban et al. 1984, Dolly and Barnard 1984). Optic lobe AChR was iodinated in vitro, immunoprecipitated with 7B2 and the precipitate analysed on an SDS gel. The three bands present, which correspond to the 48K, the heterogeneous 56K and 69K bands seen upon SDS gel electrophoresis of non-iodinated receptor, are quantitatively precipitated by 7B2 (Conti-Tronconi et al. 1985). Iodination clearly causes some breakdown of at least the 62K polypeptide – seen when the fluorographs of iodinated receptor are compared to silver- or Coomassie-stained gels of non-iodinated AChR. Since the 62K polypeptide cannot be immunoprecipitated (though possibly because of this degradation), there is no real evidence to support the contention that it forms part of a homologous AChR in optic lobe.

Direct evidence for homology to muscle AChR has been obtained by amino-terminal microsequencing of isolated subunits. The four polypeptides visualised in optic lobe AChR were isolated by preparative SDS gel electrophoresis and aliquots were rerun to show purity. In at least three separate experiments using two different receptor preparations,

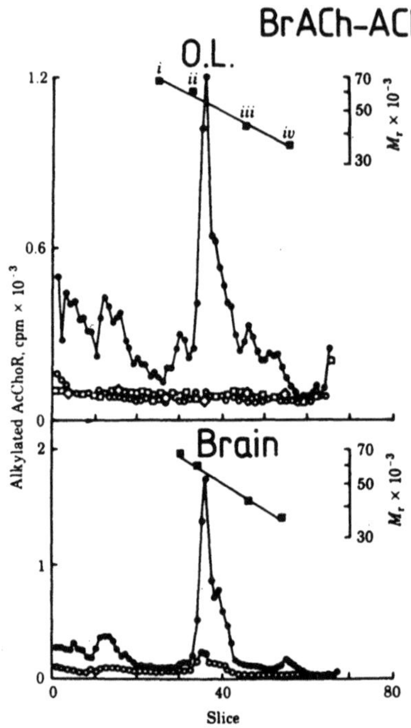


Fig. 11. Optic lobe or brain (see text) AChR was affinity labelled with bromo-[$^3$H]acetylcholine and electrophoresed on an SDS gel which was then sliced and counted. *i-iv* represent the migration of molecular weight markers. (Norman et al. 1982)

the 48K polypeptide gave a single sequence. Amino terminal microsequence (ATAS) analysis of the other three isolated polypeptides gave no signal above background, indicating that they have blocked amino termini. This blockage occurs before subunit separation, since ATAS analysis of intact optic lobe preparations (containing all four subunits) gave a clear signal which corresponds to the sequence observed for the 48K polypeptide.

Comparison of the chemically determined amino terminal amino acid sequence of the 48K polypeptide with the known amino terminal sequences of chicken muscle AChR subunits shows that it is highly homologous to the chicken muscle α-subunit (Fig. 12). The sequences can be aligned without introducing gaps and of the 26 residues available for comparison, 10 are identical. An additional five residues (shown by arrows in Fig. 12) represent chemically similar amino acid substitutions. The α-like nature of this polypeptide is further highlighted by the presence, in this brain protein, of stretches of amino acids diagnostic for the α-subunit. These residues, glu-X-lys-leu at positions 4-7 (where X can be lys or arg but no other) and arg-pro-val at positions 20-22, occur in all known α-subunit sequences, but not in beta, gamma, or delta.

In peripheral AChR, the α-subunit is known to be the subunit which is labelled by Br-ACh and binds α-BuTX. In the optic lobe AChR, we have shown that the α-like 48K subunit is not labelled, but the 56K polypeptide contains the site labelled by these reagents. Raftery and coworkers (personal communication) have noted that multiple ligand binding sites exist on *Torpedo* AChR and that Br-ACh can label other *Torpedo* subunits under different experimental conditions; thus it is possible that in the divergent CNS receptor a different subunit is more easily labelled. To summarise, the isolated optic lobe AChR appears

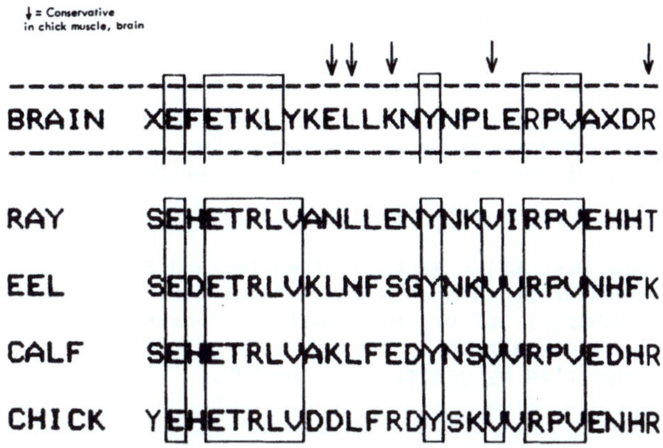


Fig. 12. The amino terminal amino acid sequence of the optic lobe 48K subunit is compared to that of the - subunits of *Torpedo*, *Electrophorus*, calf and chick. *Boxes* enclose residues identical in all five AChR's and *arrows* represent conservative amino acid substitutions when chicken muscle and brain are compared

to be an oligomeric protein which is structurally similar to, but distinctly different than the peripheral AChR: (1) the receptor is a larger complex, sedimenting at 10.1S, (2) four polypeptides of $M_r$ 48K, 56K, 62K, and 69K are consistently isolated, (3) one subunit shows distinct homology to all known α-subunits and (4) another subunit can be labelled with Br-ACh and cross-linked to α-BuTX.

## Cloning Brain AChR Genes

Experiments currently in progress aim to isolate the gene encoding the 48K polypeptide. We are using the cross-hybridisation approach – using the chicken muscle α-subunit cDNA clone as a probe to isolate homologous genes from an optic lobe cDNA library. We were, however, somewhat concerned that the loss in hybridisation signal between the (apparently) much more divergent peripheral and brain AChR might render this approach impractical. Recall the relatively weak signal observed when cross-hybridising *Torpedo* probes to the relatively conserved chicken cognate (Fig. 8). When optic lobe poly(A)$^+$ RNA is hybridised to a muscle cDNA clone on a Northern blot, no detectable signal is seen, even after a long exposure. However, hybridisation to phage plaques gives a stronger signal than can be seen in Northern or Southern blots (probably due to the higher local nucleic acid concentration on the nitrocellulose). When we screened a λgt10 optic lobe cDNA library of 200,000 clones (Huynh et al. 1985, Watson and Jackson 1985), four positively hybridising plaques were detected. This is a frequency of 0.002%, approximately the same abundance with which α-subunit sequences are present in innervated skeletal muscle mRNA and the λgt10 library prepared therefrom (see above). The concentration of α-BuTX binding sites in the two tissues is similar (Wang et al. 1978). However, it is not surprising that these clones did not hybridise to short oligonucleotide probes derived from the ATAS sequence, since they probably represent partial cDNA clones lacking the DNA sequences which encode the amino terminus of the protein. We are currently testing the presumed tissue-specific expression of these clones.

In a second approach, a single long oligonucleotide representing a "best-guess" 48-mer has been synthesised. These probes are designed to reflect the bias in codon usage for a given organism. Detection of homologous sequences in the genome tolerates a rather surprising degree of mismatch (Jaye et al. 1983, S. Anderson and Kingston 1983,

Ullrich et al. 1984a,b). This probe is being used in hybridisation experiments to a chicken genomic library.

Cloning other components of the optic lobe AChR is likely to depend either upon sequencing free N-termini of internal regions of these polypeptides revealed by limited proteolysis of intact subunits or the development of conformation-independent antibodies capable of detecting other subunits.

*Acknowledgments.* We would like to thank Drs. Brian Richards (Searle) and Michael Raftery (Caltech) for their continued interest in and support of the work described, and Dr. John Bishop (University of Edinburgh) and Dr. Marc Ballivet (Université de Genève) for helpful discussion. The work at Caltech is supported by grants from the NIH and that at Imperial College is supported by grants from the MRC.

## References

Anderson DJ, Blobel G (1981) In vitro glycosylation and membrane insertion of the four subunits of *Torpedo* acetylcholine receptor. Proc Natl Acad Sci USA 79:5598-5602

Anderson S, Kingston IB (1983) Isolation of a genomic clone for bovine pancreatic trypsin inhibitor by using a unique-sequence synthetic DNA probe. Proc Natl Acad Sci USA 80:6838-6842

Ballivet M, Patrick J, Lee J, Heinemann S (1982) Molecular cloning of cDNA coding for the gamma subunit of *Torpedo* acetylcholine receptor. Proc Natl Acad Sci USA 79:4466-4470

Ballivet M, Nef P, Stalder R, Fulpius B (1983) Genomic sequences encoding the α-subunit of acetylcholine receptor are conserved during evolution. Cold Spring Harbor Symp Quant Biol 48:83-88

Beeson DMW, Barnard EA, Conti-Tronconi B, Dunn SMJ, Anderton TL, Wilderspin AF, Bell LD, Jackson JF (1985) The chicken muscle acetylcholine receptor: subunit structure and the α-subunit cDNA cloning. J Biol Chem, in press

Betz H, Pfeiffer F (1984) Monoclonal antibodies against the α-bungarotoxin-binding protein of chick optic lobe. J Neurosci 4:2095-2105

Brockes JP, Hall ZW (1975) Acetylcholine recpetors in normal and denervated rat diaphragm muscle. II. Comparison of junctional and extrajunctional receptors. Biochemistry 14:2100-2106

Claudio T, Raftery MA (1977) Immunological comparison of acetylcholine receptors and their subunits from species of electric ray. Arch Biochem Biophys 181:481-489

Conti-Tronconi BM, Dunn SMJ, Barnard EA, Dolly JO, Lai FA, Ray N, Raftery MA (1985) Brain and muscle acetylcholine receptors are different but homologous proteins. Proc Natl Acad Sci USA 82:5208-5212

Damle V, Karlin A (1978) Affinity labelling of one of two α-neurotoxin binding sites in acetylcholine receptor from *Torpedo californica*. Biochem Biophys Res Commun 84:845-851

Devillers-Thiéry A, Giraudat J, Bentaboulet M, Changeux J-P (1983) Complete mRNA coding sequence of the acetylcholine binding α-subunit of *Torpedo marmorata* acetylcholine receptor: A model for the transmembrane organization of the polypeptide chain. Proc Natl Acad Sci USA 80:2067-2071

Dolly JO, Barnard EA (1984) Nicotinic acetylcholine receptors: an overview. Biochem Pharmacol 33:841-858

Finer-Moore J, Stroud RM (1984) Amphipathic analysis and possible formation of the ion channel in an acetylcholine receptor. Proc Natl Acad Sci USA 81:155-159

Greenberg ME, Ziff EB (1980) Stimulation of 3T3 cells induces transcription of the c-*fos* proto-oncogene. Nature (London) 311:433-437

Guy HR (1984) A structural model of the acetylcholine receptor based on partition energy and helix packing calculations. Biophys J 45:249-261

Huynh TV, Young RA, Davis RW (1985) Constructing and screening cDNA libraries in λgt10 and λgt11. In: Glover D (ed) DNA cloning: a practical approach, vol. I. IRL, Oxford, pp 49-78

Jacob MH, Berg DK (1983) The ultrastructural localization of α-bungarotoxin binding sites in relation to synapses on chick ciliary ganglion neurons. J Neurosci 3: 260-271

Jaye M, de la Salle H, Schamber F, Balland A, Kohli V, Findeli A, Tolstoshev P, Lecocq J-P (1983) Isolation of a human antihaemophilic factor IX cDNA clone using a unique 52 base synthetic oligonucleotide probe deduced from the amino acid sequence of bovine factor IX. Nucleic Acids Res 11:2325-2335

Kistler J, Stroud RM, Klymkowsky MW, Lalancette RA, Fairclough RH (1982) Structure and function of an acetylcholine receptor. Biophys J 37:371-383

Lindstrom J, Einarson B, Merlie J (1978) Immunization of rats with polypeptide chains from *Torpedo* acetylcholine receptor causes an autoimmune response to receptors in rat muscle. Proc Natl Acad Sci USA 75:769-773

Lindstrom J, Criada M, Hochschwender S, Fox L, Sarin V (1984) Immunochemical tests of acetylcholine receptor subunit models. Nature (London) 311:573-575

Marshall LM (1981) Synaptic localization of α-bungarotoxin binding which blocks nicotinic transmission at frog sympathetic neurons. Proc Natl Acad Sci USA 78: 1948-1952

Mehraban F, Kemshead JT, Dolly JO (1984) Properties of monoclonal antibodies to nicotinic acetylcholine receptors from chick muscle. Eur J Biochem 138:53-61

Mendez B, Valenzuela P, Martial JA, Baxter JD (1980) Cell-free synthesis of acetylcholine receptor polypeptides. Science 209:695

Merlie JP, Isenberg KE, Russell SD, Sanes JR (1984) Denervation supersensitivity in skeletal muscle: analysis with a cloned cDNA probe. J Cell Biol 99:332-335

Mishina M, Tobimatsu T, Imoto K, Tanaka K, Fujita Y, Fukuda K, Kurasaki M, Takahashi H, Morimoto Y, Hirose T, Inayama S, Takahashi T, Kuno M, Numa S (1985) Location of functional regions of acetylcholine receptor α-subunit by site-directed mutagenesis. Nature (London) 313:364-369

Morley BJ, Kemp GE (1981) Characterisation of a putative nicotinic acetylcholine receptor in mammalian brain. Brain Res Revs 3:81-104

Nef P, Mauron AS, Stalder C, Alliod C, Ballivet M (1984) Structure, linkage and sequence of the two genes encoding the delta and gamma subunits of the nicotinic acetylcholine receptor. Proc Natl Acad Sci USA 81:7975-7979

Noda M, Takahashi T, Tanabe T, Toyosato M, Furutani Y, Hirose T, Asai M, Inayama S, Miyata M, Numa S (1982) Primary structure of α-subunit precursor of *Torpedo californica* acetylcholine receptor deduced from cDNA sequence. Nature (London) 299: 793-797

Noda M, Takahashi H, Tanabe T, Toyosato M, Kikyotani S, Furutani Y, Hirose T, Takashima H, Inayama S, Miyata T, Numa S (1983a) Structural homology of *Torpedo californica* acetylcholine receptor subunits. Nature (London) 302:528-532

Noda M, Furutani Y, Takahashi H, Toyosato M, Tanabe T, Shimizu S, Kikyotani S, Kayano T, Hirose T, Inayama S, Numa S (1983b) Cloning and sequence analysis of calf cDNA and human genomic DNA encoding α-subunit precursor of muscle acetylcholine receptor. Nature (London) 305:818-823

Norman RI, Mehraban F, Barnard EA, Dolly JO (1982) Nicotinic acetylcholine receptor from chick optic lobe. Proc Natl Acad Sci USA 79:1321-1325

Oswald RE, Freeman JA (1981) Alpha-bungarotoxin binding and central nervous system nicotinic acetylcholine receptors. Neuroscience 6:1-14

Raftery MA, Hunkapiller MW, Strader CD, Hood LE (1980) Acetylcholine receptor: complex of homologous subunits. Science 208:1454-1457

Sumikawa K, Houghton M, Emtage JS, Richards BM, Barnard EA (1981) Active multi-subunit ACh receptor assembled by translation of heterologous RNA in *Xenopus* oocytes. Nature (London) 292:862

Sumikawa K, Houghton M, Smith JC, Bell L, Richards BM, Barnard EA (1982) The molecular cloning and characterisation of cDNA coding for the α-subunit of the acetylcholine receptor. Nucleic Acids Res 10:5809-5822

Syapin PJ, Salvaterra PM, Engelhardt JK (1982) Neuronal-like features of TE671 cells: presence of a functional nicotinic cholinergic receptor. Brain Res 231:365-377

Takai T, Noda M, Furutani Y, Takahashi H, Notake M, Shimizu S, Kayano T, Tanabe T, Tanaka K, Hirose T, Inayama S, Numa S (1984) Primary structure of gamma subunit precursor of calf-muscle acetylcholine receptor deduced from the cDNA sequence. Eur J Biochem 143:109-115

Ullrich A, Coussens L, Hayflick JS, Dull TJ, Gray A, Tam AW, Lee J, Yarden Y, Libermann TA, Schlessinger J, Downward J, Mayes ELV, Whittle N, Waterfield MD, Seeburg PH (1984a) Human epidermal growth factor receptor cDNA sequence and aberrant expression of the amplified gene in A431 epidermoid carcinoma cells. Nature (London) 309:418-425

Ullrich A, Berman CH, Dull TJ, Gray A, Lee JM (1984b) Isolation of the human insulin-like growth factor I gene using a single synthetic DNA probe. EMBO J 3:361-364

Unwin PNT, Zampighi C (1980) Structure of the junction between communicating cells. Nature (London) 283:545-549

Wang GK, Molinaro S, Schmidt J (1978) Ligand responses of α-bungarotoxin binding sites from skeletal muscle and optic lobe of the chick. J Biol Chem 253:8507-8512

Watson CJ, Jackson JF (1985) An alternative procedure for the synthesis of double-stranded cDNA for cloning in phage and plasmid vectors. In: Glover D (ed) DNA cloning: a practical approach, vol I. IRL, Oxford, pp 79-88

Wolosin JM, Lyddiatt A, Dolly JO, Barnard EA (1980) Stoichiometry of the ligand-binding sites in the acetylcholine-receptor oligomer from muscle and from electric organ. Eur J Biochem 109:495-505

Young EF, Ralston E, Blake J, Rachmandran J, Hall ZW, Stroud RM (1985) Topological mapping of the acetylcholine receptor: Evidence for a model with five transmembrane segments and a cytoplasmic COOH-terminal peptide. Proc Natl Acad Sci USA 82:626-630

# Laser-Flash Photoaffinity Labeling of Acetylcholine Receptors with Millisecond Time Resolution

A. Fahr[1], S. Hellmann[2], L. Lauffer[3], P. Muhn[4], and F. Hucho[4]

## Introduction

The function of membrane-bound receptors is the transduction of signals from the outside of a cell to the inside. The signal itself and therefore the transduction process, too, are transient molecular events. In the case of the nicotinic acetylcholine receptor (nAChR) the signal consists of acetylcholine molecules binding very rapidly (diffusion controlled, Maelicke 1984) to the receptor protein. The transduction process consists of a change of the electrical properties of the receptor membrane. It is based on ion currents through transiently opened ion channels formed by the receptor protein (Hille 1984).

In contrast to the abundance of information available concerning the primary and quaternary structure of the nAChR (Popot and Changeux 1984), there is relatively little known about the mechanism of its activity. Especially the mechanism of coupling between signal recognition (acetylcholine binding) and channel opening, and moreover the identity and mechanism of functioning of the channel are largely unknown (Changeux 1981). The kinetic methods employed so far in these investigations such as, e.g., measurements of intrinsic or extrinsic fluorescence changes, quenched ionflow and patchclamp recordings of single channel currents (Changeux 1981, Hess et al. 1983) yield little information as to the molecular structures involved in the underlying processes. Rather vaguely conformational changes of the receptor protein are postulated. Here we propose a method allowing labeling of irreversibly transient receptor states with a time resolution in the millisecond range. We believe that this method will enable us to identify specific sites of the receptor protein which are directly involved in the gating and regulation of the receptor ion channel.

## $TPMP^+$ (Triphenylmethylphosphonium) as a Specific Ion Channel Blocker

The lipophilic cation has been shown previously to be a non-competitive antagonist of the nAChR of *Torpedo marmorata*.

[1] Institut für Atom- und Festkörperphysik, Abtlg. Biophysik, Freie Universität Berlin, Arnimallee 14, D-1000 Berlin 33

[2] Present address: Max-Planck-Institut für Ernährungsphysiologie, Rheinlanddamm 201, D-4600 Dortmund, FRG

[3] Present address: UCSF, Medical School, Dept. of Biochemistry and Biophysics, San Francisco, CA 94143, USA

[4] Institut für Biochemie, Arbeitsgruppe Neurochemie, Freie Universität Berlin, Thielallee 63, D-1000 Berlin 33

36. Colloquium - Mosbach 1985
Neurobiochemistry

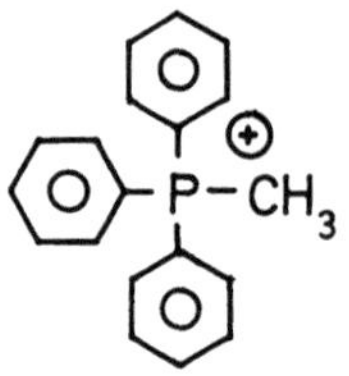

$TPMP^+$

Binding studies have shown that up to a concentration of about 0.1 mM it does not displace $^3$H-acetylcholine to a significant extent from the agonist binding site. Aldready at 0.01 mM $TPMP^+$ blocks completely the ion flux from *Torpedo* receptor-rich membrane vesicles (microsacs). $TPMP^+$ competes on the other hand with well documented channel blockers such as, e.g., $^3$H-PCP (phencyclidine), local anesthetics and the non-ionic detergent Triton X-100 (Lauffer and Hucho 1982).

The affinity of $TPMP^+$ for *Torpedo* AChR as measured by direct binding and by competition studies with $^3$H-PCP was shown to be enhanced by agonists ($K_D$ = 13 μM in the absence, $K_D$ = 1.5 μM in the presence of 10 μM carbamoylcholine). The stoichiometry was one $^3$H-$TPMP^+$ bound/ two acetylcholine binding sites. Electrophysiological experiments with the frog neuromuscular junction confirmed the results obtained with *Torpedo* microsacs (Spivak and Albuquerque 1985).

## Inhibitory Effects of $TPMP^+$ at the Frog End-Plate

The effect of $TPMP^+$ on synaptic transmission has been studied in the cutaneous pectoris muscle of the frog: the animal was decerebrated, demedullated, then the muscle was removed from the animal and pinned down on silicone rubber in an experimental chamber. The endplate areas were freed from connective tissue using watchmaker forceps. The cut end of the nerve was then sucked into a suction pipette and stimulated continuously at 1 Hz with 0.2 ms pulses delivered from pulse generator. While isolation of the muscle was performed in Ringer solution: (in mM), NaCL: 115, KCl: 2, $CaCl_2$: 1.8, and Hepes buffer: 5 (pH = 7), the experiments were made in a solution obtained by mixing the above solution with another one where $Ca^{2+}$ was replaced by $Mg^{2+}$ (5 mM) to avoid contractions. The effective Ca/Mg ratio was usually around 1:4. Muscle fibers were impaled with two microelectrodes and their membrane potential voltage clamped. Temperature was kept below 20°C by cooling the perfusion fluid. $TPMP^+$ was applied by diluting a $10^{-2}$ M stock Ringer solution into the bath solution. Before starting measurements of the endplate current (epc) evoked by nerve stimulation, the preparation was equilibrated with each drug concentration for about 3 minutes; epc's were measured for one minute at any membrane potential and the average of at least 30 epc's calculated; values in the figures have been collected from 8 fibers out of 4 muscles.

*Resting Membrane Potential.* $TPMP^+$ had no effect on the resting potential nor the membrane resistance of the muscle cells at the concentrations tested ($10^{-6}$ M to $3 \times 10^{-5}$ M).

*Inhibitory Action on Postsynaptic Currents.* At a given potential (e.g., -90 mV, ■ in Fig. 1) $TPMP^+$ inhibited the evoked response in a dose-dependent manner: the $ID_{50}$ was close to $10^{-5}$ M. However, the inhibition varied with membrane potential: depolarizing the membrane poten-

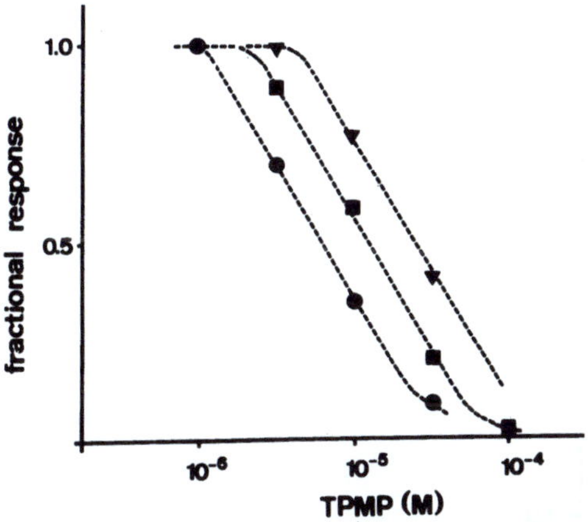


Fig. 1. Relative inhibition of the epc at the frog endplate produced by different concentrations of $TPMP^+$ at membrane potentials of -60 mV (▲), -90 mV (■), and -120 mV (●)

tial by 30 mV increased the $ID_{50}$ by approximately 1/4 of a log unit; a 30 mV hyperpolarization had the opposite effect (▲ and ● in Fig. 1).

The sensitivity to the inhibitory action of $TPMP^+$ on postsynaptic currents is further described in Fig. 2. Control peak epc measurements were performed after $TPMP^+$ tests (open circles). They are plotted against the membrane potential relatively to the value obtained at -80 mV. In the presence of $TPMP^+$ ($10^{-5}$ M) the peak amplitude of the epc is reduced and the voltage sensitivity is modified: hyperpolarizing the cells stepwise from -50 mV, at which voltage there is practically no inhibition, the amplitude of the relative epc's first increases, although less than in control epc's, then decreases, resulting in a bell-shaped curve. Depolarization of the membrane from -140 mV first leads to a slight increase of the amplitude, thereafter it is practically constant but below the values obtained during the hyperpolarization run. Around -70 mV, both curves superimpose on each other. A similar potential dependence seems to exist at other $TPMP^+$ concentrations: at $3 \times 10^{-5}$M, inhibition is stronger and the bell-shaped response curve seems to be shifted to more depolarized membrane potentials. Conversely, for lower concentrations of $TPMP^+$ it would be expected to be shifted to more hyperpolarized membrane potentials.

When the test epc's are compared to control values obtained before $TPMP^+$ applications, inhibition seems to be less strong. In other words, $TPMP^+$ seems to induce a long-lasting potentiation of the control epc's (see inset in Fig. 2). It is not known if the origin of this effect is pre- or postsynaptic; however, no long-lasting action of $TPMP^+$ on miniature endplate potential frequency nor amplitude was noticed.

*Conclusion.* These results show that $TPMP^+$ inhibits the acetylcholine action at the frog endplate and that its effect is strongly sensitive to the membrane potential. On the basis of available evidence and current interpretation it may be concluded that its inhibitory action is probably not due to a competition with ACh at the receptor site but results from an interaction at some other site(s) sensitive to the membrane potential. Recently these findings have been confirmed and extended by Spivak and Albuquerque (1985), who found in addition to the voltage dependence a shortening of the channel lifetime by $TPMP^+$.

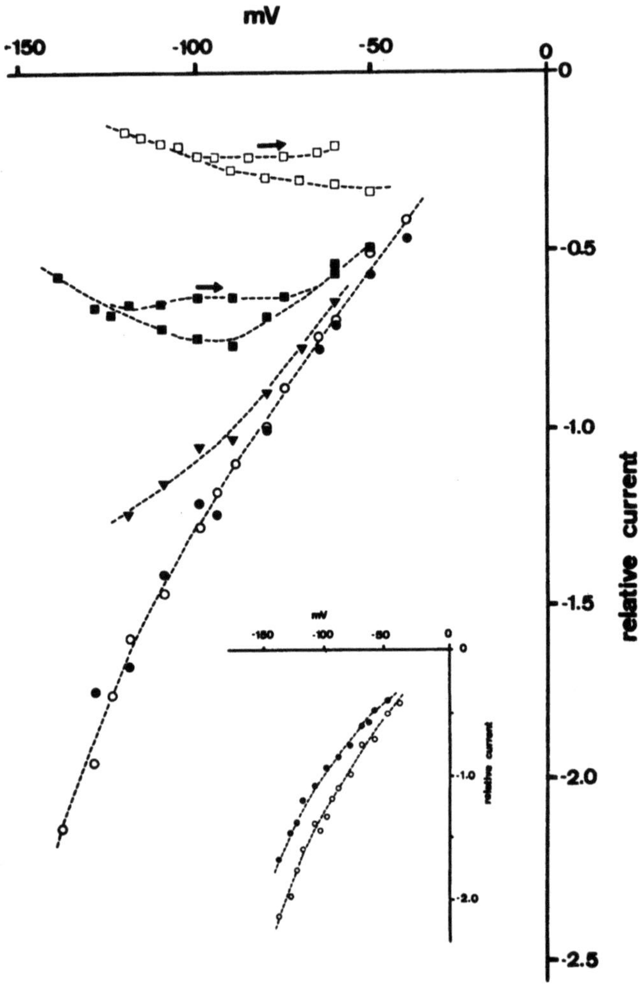


Fig. 2. Voltage sensitivity of the epc recorded at the frog endplate in the presence of $TPMP^+$ at 1 µM (●), 3 µM (▲), 10 µM (■), and 30 µM (□). Control values (o) were collected after washing out the drug. (*Inset* compares the voltage sensitivity of epc's before (●) and after (o) the application of 10 µM $TPMP^+$ for 15 min). The values were obtained by stepwise byperpolarizing the membrane. Stepwise depolarization is indicated by *arrows*. Epc's are derived from several fibers and are expressed relative to the value obtained at -80 mV (actual) currents ranged from -30 nA to -100 nA)

## Covalent Attachment of $TPMP^+$ to Its Site of Action

The aim of our biochemical studies is to identify the ion channel of the AChR within the receptor protein complex or at least an allosteric site regulating its gating. $TPMP^+$ may be a useful tool for this purpose because it can be used as a photoaffinity label: Irradiation of an AChR-$^3H$ $TPMP^+$ complex with UV light causes – though in low yields – irreversible incorporation of radioactivity into the receptor (Muhn and Hucho 1983). Expecially with AChR from *Torpedo marmorata*, the site of incorporation depends on the functional state of the receptor. Agonists and competitive antagonists (Muhn et al. 1983) [as well as treatment of the receptor-rich membranes with phospholipase $A_2$ or "spontaneous" aging of the membranes (Muhn et al. 1983, Muhn et al. 1984b)] cause a shift of the predominant site of the photoreaction from the $\alpha$- to the $\delta$-polypeptide chain (and to a lesser extent to the $\beta$-chain). To exploit this shift for the biochemical (i.e., protein chemical) analysis of transient receptor states, we developed a method for *time-dependent photoaffinity labeling with millisecond time resolution* (Muhn et al. 1984a,b). For this we combined a stopped flow apparatus with a high energy pulse laser. Figure 3 shows a scheme of the experimental setup. Its special feature (besides an optical circuitry to measure the velocity of the syringe pistons for estimating the dead time of the stopped-flow apparatus) is a modified mixing cell shown in more detail in Fig. 4. Its mixing chamber has a cylindrical shape (1.7 mm diameter, 1.5 mm in length) and is connected to the reaction chamber by a gap (1.4 mm × 3 mm) between the cell and the quartz glass plate covering the reaction chamber on both sides.

The reaction chamber is a circular cavity, 2.4 mm in diameter and 5.5 mm in length. The exit channel of the cell is essentially a hole (1 mm in diameter) at which a teflon tube was fixed by means of a two-component glue. Through it the reaction products can be collected after photolabeling.

The dead time of the apparatus was 2.4 ms. Samples of 30-35 µl were removed after each stopped flow experiment using a Hamilton syringe.

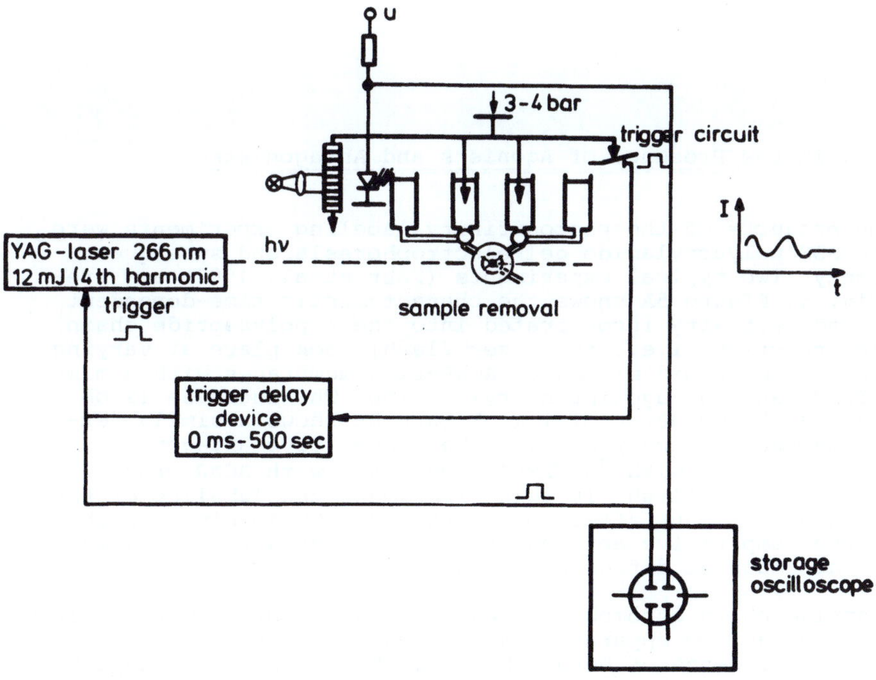


Fig. 3. Schematic drawing of the main components of the rapid laser flash photolabeling setup

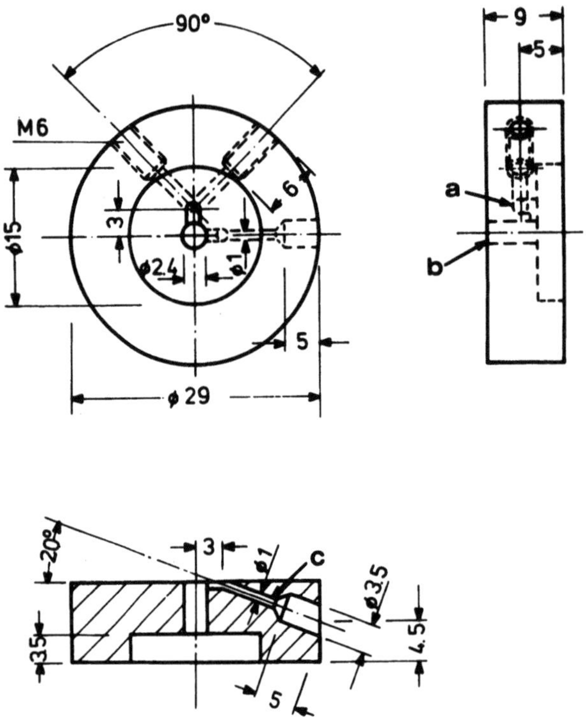


Fig. 4. Detailed drawing of the metal mixing cell of the stopped-flow apparatus. a mixing chamber, b reaction chamber, c exit channel. Both sides of the reaction chamber are covered by quartz glass plates

## Photolabeling in the Presence of Agonists and Antagonists

The reaction products of the photoaffinity labeling experiments were subjected to SDS polyacrylamide gel electrophoresis and subsequent autoradiography. Two typical experiments (Fahr et al. 1985) are presented in Fig. 5. Figure 5A shows the characteristic time-dependent increase of radioactivity incorporated into the δ-polypeptide chain when the photoreaction (i.e., the laser flash) took place at varying time intervals after rapid mixing of AChR-rich membranes with a mixture of $^3$H-TPMP$^+$ and the agonist acetylcholine. The increase is obvious in trace 3, 50 ms after mixing. Figure 5B shows a similar experiment but instead of acetylcholine the competitive antagonist hexamethonium (together with $^3$H-TPMP$^+$) was mixed with AChR prior to triggering to laser flash. In this experiment the labeling of the δ-chain became significant after 50 s (trace 8). Obviously both the agonist and the competitive antagonist induced a structural change in the protein, although at different rates.

The α-polypeptide chain becomes labeled immediately after mixing (dead time of the stopped-flow apparatus was 2.4 ms) in both cases. Interestingly, in the presence of hexamethonium, the peripheral receptor-associated α'-peptide is labeled as well. The α'-peptide is located on the inside of the AChR-rich vesicular membrane of the receptor com-

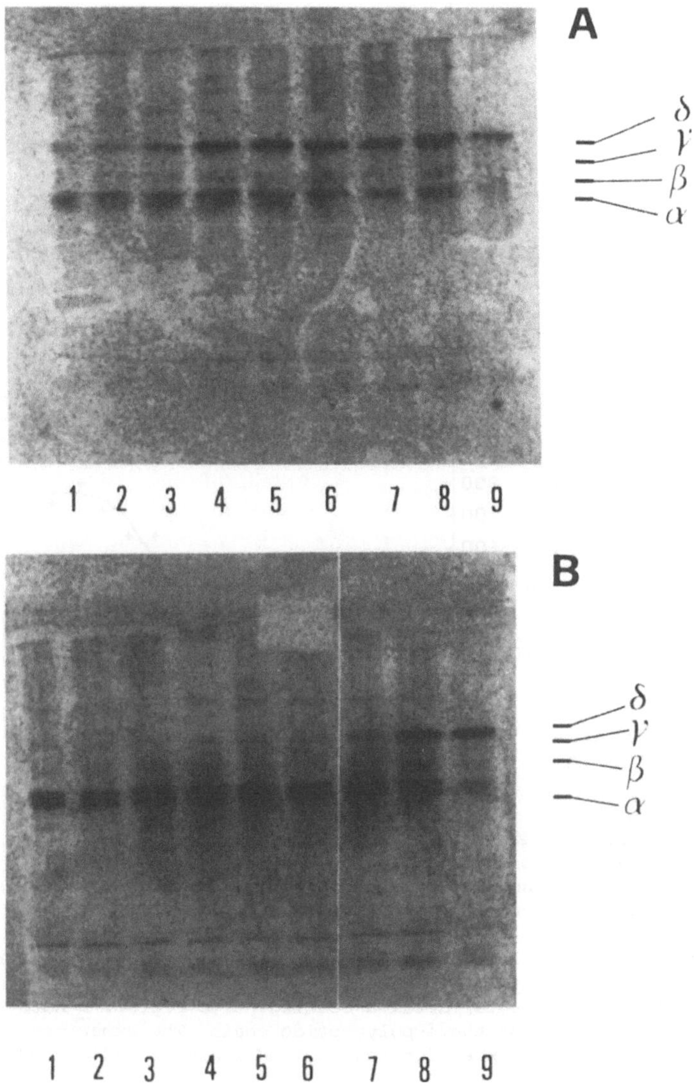


Fig. 5 A,B. Time-resolved photoaffinity labeling of AChR stimulated by an agonist (A) or a competitive antagonist (B). The autoradiograms of SDS-polyacrylamide gel electrophoresis show AChR-polypeptide chains photoaffinity-labeled with the noncompetitive antagonist ($^3$H)TPMP$^+$ at different time lapses after mixing AChR-rich membranes with a mixture of ($^3$H)TPMP$^+$ and acetylcholine or hexamethonium, respectively. After mixing in a stopped-flow apparatus the laser was triggered after *1* 8 ms; *2* 23 ms; *3* 50 ms; *4* 200 ms; *5* 500 ms; *6* 2 s; *7* 5 s; *8* 50 s; *9* 330 s. In the experiment B the time lapses were identical to those in A. In the stopped-flow apparatus one syringe contained receptor-rich membranes (1 mg/ml) in Ringer solution; the other syringe contained ($^3$H) TPMP$^+$ (0.4 μM) and acetylcholine (A) or hexamethonium (B), respectively, at 0.1 mM. The irradiated volume was 22 μl, the dead time of the stopped-flow apparatus was 2.4 ms, irradiation time (duration of 1 pulse of the laser) was about 4 ns with a pulse energy of 15 m at 266 nm wavelength. The experiment was performed at room temperature. After irradiation the sample was removed from the mixing chamber and analyzed by SDS-polyacrylamide gel electrophoresis

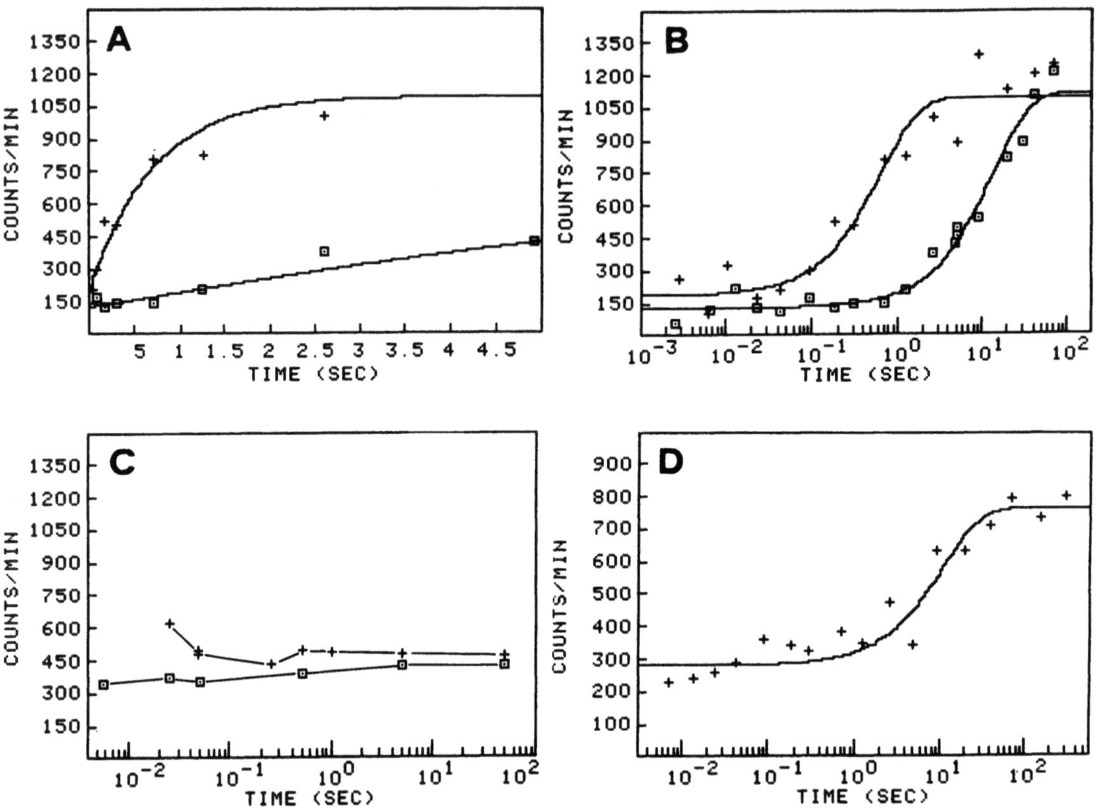


Fig. 6 A-D. Time-resolved photoaffinity labeling of AChR with $(^3H)TPMP^+$ stimulated by cholinergic effectors. A, B Labeling of the δ-polypeptide chain; C labeling of the α-polypeptide chain. (+) 0.1 mM carbamoylcholine added to receptor-rich membranes simultaneously with the photolabel; (□) receptor-rich membranes preincubated with 0.1 mM carbamoylcholine for about 10 min, then mixed in the stopped-flow apparatus with $(^3H)TPMP^+$/carbamoylcholine. A and B show the same experiment plotted with linear and with logarithmic time scale, respectively, to visualize the kinetics in the different time ranges. For the other experimental details see legend to Fig. 1. D Antagonist-stimulated photoaffinity labeling of the δ-polypeptide chain. The experimental conditions were as in Fig. 5 with the exception that 0.1 mM flaxedil and 0.7 μM $(^3H)TPMP^+$ were used

plex. Its labeling therefore is not surprising, since $TPMP^+$ is a lipophilic cation known to be able to cross membranes.

In order to evaluate the results in a more quantitative manner, in the following experiments we excised the bands of the Coomassie-stained gels corresponding to the α- and δ-polypeptide chains and determined their radioactivity by liquid scintillation counting. Figure 6 shows a series of time-resolved photoaffinity labeling experiments. A considerable difference in the kinetics of the labeling of the δ-polypeptide chain was observed depending on the mode of stimulation by an agonist (in this case by 0.1 mM carbamoylcholine): Carbamoylcholine, if mixed simultaneously with AChR-rich membranes and $(^3H)TPMP^+$ (Fig. 6 A,b), promotes the δ-reactive state in a similar manner as acetylcholine (completion of the reaction after about 10 s, $t_{1/2} = 450$ ms). In contrast, after preincubation with carbamoylcholine, a different kinetic pattern emerges (completion after about 100 s, $t_{1/2} = 9.5$ s) (Fig. 6 A,B). As observed before (cf. Fig. 5), the amount of label incorporated in the α-polypeptide chain remains more or less the same or decreases slightly (Fig. 6 C).

Competitive antagonists also stimulate the photoreaction with the δ-polypeptide chain. With all antagonists tested, i.e., hexamethonium (Fig. 5 B), flaxedil (Fig. 6 D), and d-tubocurarine, we found no fast stimulation of δ-labeling irrespective if the membrane suspension was mixed simultaneously with antagonist and ($^3$H)TPMP$^+$ or had been preincubated with antagonist before mixing. The kinetics in the presence of antagonists differ slightly from the one obtained after preincubation with agonists. The calculated time constant of the former is 11 s of the latter 14 s (carbamoylcholine).

The experimental data were evaluated assuming a single exponential rate equation. Evaluation with two exponentials would fit the data only slightly better. Therefore with the number of data points available we cannot decide whether the reaction is (pseudo-)first order or bimolecular. In our experiments, concentration of bungarotoxin binding sites were chosen to be about 20-fold in excess over the ($^3$H)TPMP$^+$ concentration to favor a pseudo-first-order reaction mechanism.

Rapid mixing of AChR-rich membranes with ($^3$H)TPMP$^+$ alone and laser-flash photolabeling at different times after mixing resulted in no significant changes of the labeling pattern. Similarly preincubation of membranes with antagonist prior to rapid mixing with ($^3$H)TPMP$^+$ and time-dependent photolabeling yielded no unequivocal kinetics.

*Conclusion.* Rapid photoaffinity labeling of AChR has been proposed recently by several laboratories (Muhn et al. 1984a, Cox et al. 1984, Heidmann and Changeux 1984). Only the laser method provides a time resolution comparable to physiological events. The kinetic analysis presented here shows that at least three receptor states can be distinguished: either (a) the α-chain is the target of the photoreaction, or (b) the δ-chain becomes accessible with a $t_{1/2}$ = 500 ms, or (c) finally the δ-chain becomes accessible with a $t_{1/2}$ of the order of seconds. State (a) comprises the resting and the activated (channel open) state, state (b) represents an agonist-induced state, and state (c) is induced by competitive antagonists and by desensitizing conditions. By choosing the appropriate time scale, and in combination with finger print analysis, detailed insights into dynamic processes involving the ion channel of the receptor are to be expected.

*Acknowledgment.* This work was supported by the Deutsche Forschungsgemeinschaft and the Fonds der Chemischen Industrie. The electrophysiological experiments were performed under the guidance of Dr. Stinnakre, CNRS Laboratoire de Neurobiologie Cellulaire, Gif-sur-Yvette, France.

## References

Changeux J-P (1981) The acetylcholine receptor: An "allosteric" membrane protein. Harvey Lect 75:85-254

Cox RN, Kaldany R-R, Brandt PW, Ferren B, Hudson RA, Karlin A (1984) A continuous flow, rapid mixing, photolabeling technique applied to the acetylcholine receptor. Anal Biochem 136:476-486

Fahr A, Lauffer L, Schmidt D, Heyn M, Hucho F (1985) Covalent labeling of functional states of the acetylcholine receptor. Eur J Biochem 147:483-487

Heidmann T, Changeux J-P (1984) Time-resolved photolabeling by the noncompetitive blocker chlorpromazine of the acetylcholine receptor in its transiently open and closed ion channel conformations. Proc Natl Acad Sci USA 81:1897-1901

Hess GP, Cash DJ, Aoshima H (1983) Acetylcholine receptor-controlled ion translocation. Annu Rev Biophys Bioeng 12:443-473

Hille B (1984) Ionic channels of excitable membranes. Sinauer Assoc Inc, Sunderland, Mass, USA

Lauffer L, Hucho F (1982) Triphenylmethylphosphonium is an ion channel ligand of the nicotinic acetylcholine receptor. Proc Natl Acad Sci USA 79:2406-2409

Maelicke A (1984) Biochemical aspects of cholinergic excitation. Angew Chem Int Ed Engl 23:195-221

Muhn P, Hucho F (1983) Covalent labeling of the acetylcholine receptor from *Torpedo* electric tissue with the channel blocker ($^3$H) triphenylmethylphosphonium by ultraviolet irradiation. Biochemistry 22:421-425

Muhn P, Lauffer L, Hucho F (1983) The acetylcholine receptor and its ion channel. In: Sund H, Veeger C (eds) Mobility and recognition in cell biology. De Gruyter, Berlin

Muhn P, Fahr A, Hucho F (1984a) Photoaffinity labeling of acetylcholine receptor in millisecond time scale. FEBS Lett 166:146-150

Muhn P, Fahr A, Hucho F (1984b) Rapid laser flash photoaffinity labeling of binding sites for a noncompetitive inhibitor of the acetylcholine receptor. Biochemistry 23:2725-2730

Popot JJ, Changeux JP (1984) Nicotinic receptor of acetylcholine: Structure of an oligomeric integral membrane protein. Physiol Rev 64:1162-1222

Spivak LE, Albuquerque EX (1985) Triphenylmethylphosphonium blocks the nicotinic acetylcholine receptor noncompetitively. Mol Pharmacol 27:246-255

# The Glycine Receptor

H. Betz, B. Schmitt, G. Grenningloh, C.-M. Becker, and I. Hermans[1]

## Introduction

In the vertebrate CNS, many different compounds including acetylcholine, catecholamines, peptides, and amino acids have been identified as bona fide or candidate neurotransmitter substances. Amongst these the amino acids glycine and GABA have been particularly well characterized. These two amino acids are the most important mediators of inhibitory responses in the brain and spinal cord of all vertebrates. Both glycine and GABA produce an increase of the chloride conductance of the neuronal membrane, i.e., they hyperpolarize the postsynaptic cell. The receptors for these amino acids have been investigated in detail by anatomical, electrophysiological, and pharmacological techniques, and are now thought to be separate molecules on the basis of the following criteria:

1. Glycine and GABA receptors are differently distributed in the CNS. Glycine is the major neurotransmitter in spinal cord and brain stem, whereas GABA predominates in the brain (Snyder and Bennett 1976);
2. although some amino acids, such as taurine, can activate both the GABA and the glycine receptor, the pharmacology of the two receptors is clearly different (see below);
3. photoaffinity labeling experiments have shown that the ligand-binding polypeptides of the glycine receptor and the GABA receptor have different molecular weights (Graham et al. 1983).

Despite these topological, pharmacological, and biochemical differences, the glycine- and GABA-activated chloride channels show striking similarities in their conductance properties (Barker and McBurney 1979, Hamill et al. 1983). Therefore, a common ion-channel component for this class of receptor proteins has recently been proposed (Hamill et al. 1983). Since 1980, our laboratory has focused on the structural and immunological investigation of one of these receptors, the glycine receptor protein. Here, we discuss the currently available data and suggest a structural model for this chemically gated neuronal ion channel. In addition, data concerning the glycine receptor deficiency of the mutant mouse *spastic* are summarized.

## Pharmacology of the Glycine Receptor

The number of currently available glycine receptor ligands is comparatively small. Besides glycine, the amino acids β-alanine and taurine are known to be efficient agonists of this receptor ($K_D \sim 5-20\ \mu M$; Pfeiffer et al. 1982); GABA, in contrast, has little effect below 1 mM.

[1] Institut für Neurobiologie, ZMBH, Zentrum für Molekulare Biologie der Universität Heidelberg, Im Neuenheimer Feld 364, D-6900 Heidelberg, FRG

36. Colloquium - Mosbach 1985
Neurobiochemistry

Strychnine, a heterocyclic alkaloid from *Strychnos nux vomica*, is an extremely potent antagonist of the glycine receptor ($K_D$ 5-10 nM). All toxic effects of this alkaloid can be attributed to blocking glycinergic transmission in different regions of the CNS (Zarbin et al. 1981). In 1973, Young and Snyder demonstrated high-affinity binding of ($^3$H)-strychnine to membrane preparations of rat spinal cord. This binding was inhibited not only by glycine and other agonists, but also by chloride and related anions capable of mediating glycine responses. Strychnine binding therefore has been postulated to occur at a site associated with the ion channel of the glycine receptor (Young and Snyder 1974). Different investigators have used ($^3$H)strychnine binding to quantitate glycine receptors in a simple radio-receptor assay. We have found strychnine and its derivatives powerful tools in analyzing and purifying the glycine receptor protein.

Avermectin B1a, an anthelmintic macrocyclic lactone disaccharide, is known to increase GABA responses by binding to an allosteric site of the GABA receptor (Fritz et al. 1979). Interestingly, this compound is a potent inhibitor ($K_D < 1$ μM) of ($^3$H)strychnine binding to spinal cord membranes or detergent extracts (Graham et al. 1982). This observation supports the previously noted similarity between glycine and GABA-receptor proteins.

## Biochemistry of the Glycine Receptor

The first step in the biochemical characterization of the glycine receptor was the solubilization of ($^3$H)strychnine binding sites from synaptic membrane fractions of rat spinal cord (Pfeiffer and Betz 1981). These experiments showed that this receptor is a glycoprotein of $M_r \sim$ 250,000 whose native conformation stringently depends on the presence of phospholipids. Experiments using the solubilized glycine receptor thus have to be performed in buffers containing mixed phosphatidylcholine/detergent micelles.

The further investigation of the glycine receptor was greatly aided by the discovery that strychnine is a natural photoaffinity ligand of this receptor protein (Graham et al. 1981). Upon UV-illumination in the presence of spinal cord membranes, ($^3$H)strychnine was irreversibly incorporated into its high-affinity receptor binding site. SDS-polyacrylamide gel electrophoresis of the illuminated membranes identified a polypeptide of $M_r = 48{,}000$ (48 K) as the strychnine-binding subunit of the glycine receptor. The kinetics and the pharmacology of the UV-induced incorporation of ($^3$H)strychnine into membranes are consistent with the hypothesis that strychnine is cross-linked at or close to the antagonist binding site of GlyR (Graham et al. 1983). Also, this photoaffinity labeling proved useful for exploring the membrane topology of the 48 K subunit. Incubation of ($^3$H)strychnine-labeled membranes with different proteases led to a degradation of the 48 K band to polypeptides of lower molecular weight. Even extensive proteolysis failed, however, to remove the radioactive label from the membranes. Strychnine binding is therefore thought to occur in a receptor domain which is protected by the hydrophobic environment of the lipid bilayer (Graham et al. 1983).

An important class of tranquilizer drugs, the benzodiazepines, affect GABA-ergic transmission and can be incorporated into a membrane polypeptide of $\sim$50 K upon UV illumination (Möhler et al. 1980). This GABA-receptor subunit is different from the strychnine binding polypeptide of the glycine receptor (Graham et al. 1983).

For purification of the glycine receptor, a physiologically active derivative of strychnine, 2-aminostrychnine, was coupled to agarose beads via a long hydrophilic spacer (Pfeiffer et al. 1982). This affinity resin bound 80%-90% of the ($^3$H)strychnine-binding sites present in detergent extracts of spinal cord membranes. Furthermore, about 20% of the glycine receptor was recovered after washing of the resin and biospecific elution with the receptor agonist glycine. This protocol provides a simple procedure for isolating GlyR to near homogeneity in one chromatographic step: Currently, 50-100 μg of purified receptor protein can be obtained within 2-3 days. Also, this affinity-purification procedure could be successfully employed for the purification of the glycine receptor of pig and mouse spinal cord (Graham et al. 1985; C.-M. Becker, unpublished data).

Upon SDS-polyacrylamide gel electrophoresis, purified GlyR from rat (or mouse, or pig) spinal cord is separated into three polypeptides of 48 K, 58 K, and 93 K (Fig. 1). The stoichiometry of these polypeptides has not been clarified; however, the 48 K polypeptide is usually more abundant than the other two. The 48 K and 58 K polypeptides bind concanavalin A and thus are glycoproteins (Graham et al. 1985).

UV-illumination of the purified glycine receptor in the presence of ($^3$H)strychnine radiolabeled the 48 K and, to a much lesser extent, the 58 K polypeptide. These polypeptides thus participate in the formation of the extracellular ligand-binding domain of the glycine receptor. The function of the 93 K polypeptide is so far unknown. Recent experiments have shown that, under certain conditions of gradient centrifugation, it can be separated from the 48 K and 58 K glycine receptor subunits. Also, the 93 K polypeptide is rapidly lost in the absence of protease inhibitors (Graham et al. 1985). An interesting speculation relates the 93 K polypeptide to the ion-channel function of GlyR. Its loose receptor association and its intracellular exposition (see below) are consistent with this hypothesis. Alternatively, this polypeptide may have other functions in the postsynaptic membrane.

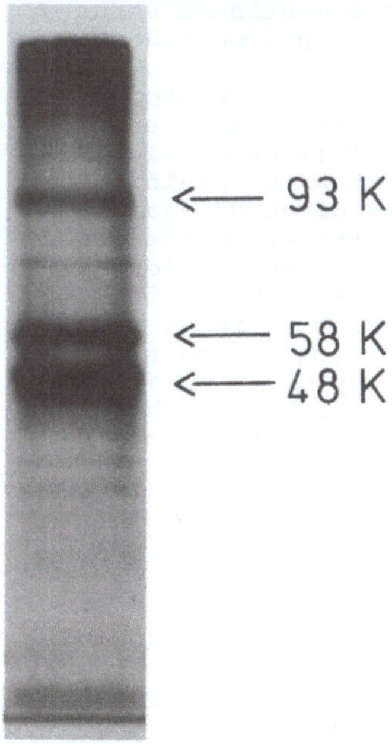


Fig. 1. Polypeptide composition of the affinity-purified glycine receptor of rat spinal cord. *Arrows* indicate the different glycine receptor polypeptides

Peptide-mapping experiments of $^{125}$I-labeled glycine-receptor polypeptides produced some peptide fragments of identical size from all three receptor subunits. The 48 K and 58 K polypeptides in particular gave rather similar proteolytic fingerprints (Pfeiffer et al. 1984). These results suggest that the different subunits of the GlyR may contain homologous domains and thus evolved from a common ancestor polypeptide, as has been shown for the nicotinic acetylcholine receptor of fish electric organ (Noda et al. 1983).

In preliminary experiments, affinity-purified glycine receptor was shown to produce glycine-induced conductance changes upon reconstitution into planar lipid bilayers (Graham and Schindler, unpublished experiments). These conductance changes were glycine-specific, strychnine-sensitive, and chloride-dependent. It therefore appears that affinity-purified glycine receptor contains both the ligand-binding and ion-transport functions of this membrane-bound signal-transducer protein.

## Immunology of the Glycine Receptor

In order to develop new ligands for different functional sites of the glycine receptor, we produced polyclonal and monoclonal antibodies against the affinity-purified receptor protein. Using standard procedures of hybridoma production and soft-agar cloning, nine different cell lines were established which secrete monoclonal antibodies (mAbs) capable of binding the $^{125}$I-labeled glycine receptor (Pfeiffer et al. 1984). In Western blots, four of these antibodies bound to one of the different denatured receptor subunits; three mAbs did not bind to blots and are thus presumably conformation-dependent. Most interestingly, however, two of the antibodies recognized more than one of the glycine receptor polypeptides: mAb 4a bound to the 48 K and 58 K polypeptides and mAb 7a to the 93 K and, to a lesser extent, the 48 K polypeptides. These immunological data extend the above-mentioned structural similarities of the different glycine receptor polypeptides.

In collaboration with other groups, some of our mAbs were tested in immunocytochemical experiments on primary cultures and tissue sections of spinal cord and brain stem (St. John et al. 1984, Triller et al. 1985). These preliminary experiments revealed (a) a co-extensive staining of synaptic regions and neuronal processes by 3 mAb's directed against different glycine receptor polypeptides; (b) differences in the accessibility of the binding domains of the mAb's specific for different subunits of the receptor. Antibodies recognizing the 48 K and 58 K polypeptides stained intact cultured neurons, whereas staining with mAbs specific for the 93 K polypeptide was observed only with sectioned material. Furthermore, by electron microscopy the antibodies were shown to bind to the inner leaflet of the neuronal membrane. These experiments establish the glycine receptor as a transmembrane protein.

## A Structural Model of the Glycine Receptor

On the basis of our previous data, we have proposed a structural model of the glycine receptor as shown in Fig. 2 (Betz et al. 1983). According to this working hypothesis, the glycine receptor is a heteropolymer of two copies of the 48 K polypeptide and one copy each of the 58 K

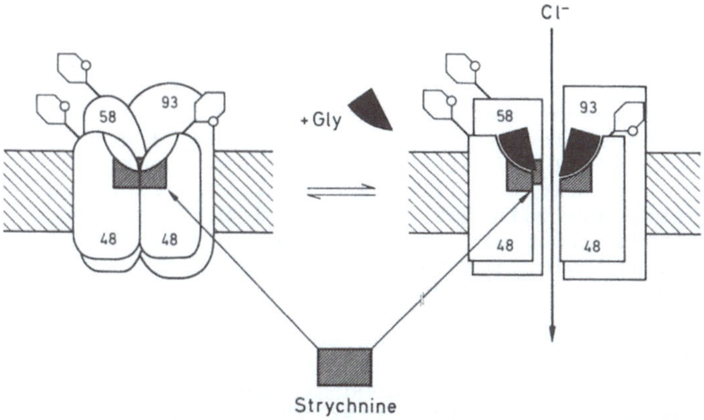


Fig. 2. A hypothetical model of the glycine receptor in its closed and open conformations

and the 93 K polypeptides. From competition experiments, the agonist-binding sites of the receptor were assigned to the 48 K polypeptide chains. In agreement with this proposal, electrophysiological data suggest that more than one glycine molecule is required for chloride-channel opening (Hamill et al. 1983). The number of strychnine-binding sites is not known; however, our data are consistent with a single strychnine-binding site which overlaps (one of) the agonist site(s) and includes not only the (two) 48 K chain(s), but also the 58 K polypeptide.

So far this model of the GlyR has not been proved. However, certain of its features can now be refined: (1) Our antibody-binding experiments suggest that the major portions of the 48 K and the 58 K polypeptides are extracellular; also, these polypeptides have been shown to be glycosylated; (2) the 93 K polypeptide differs from the other glycine-receptor subunits in that it does not bind concanavalin A and is accessible for mAb binding from the cytoplasmic site of the membrane. Whether the 93 K polypeptide has a function in ion translocation remains to be determined.

It should be noted here that the structural model proposed for the glycine receptor may also apply to the GABA-receptor protein (Sigel and Barnard 1984): (1) Upon affinity chromatography, two GABA-receptor polypeptides of 53 K and 57 K were identified which are both labeled by benzodiazepines; (2) from irradiation inactivation studies a GABA receptor polypeptide of >100 K has been suggested. In view of the astonishing similarity of glycine- and GABA-mediated chloride-conductance changes, considerable homology between these two receptor systems can be expected. The molecular cloning and sequencing of cDNA coding for the different glycine and GABA-receptor polypeptides will provide an answer to these questions. As a first step toward these goals, the N-terminal region and an internal peptide of the 48 K glycine receptor polypeptide were microsequenced (collaboration with K. Beyreuther, University of Cologne). Oligonucleotide mixtures corresponding to the available amino-acid sequences are currently being used to screen cDNA libraries of rat spinal cord mRNA established in bacteriophage λgt 10 (Grenningloh et al., unpublished experiments).

## Glycine Receptor Deficiency in the Mutant Mouse *Spastic*

Neurological mutants of the mouse, *Mus musculus*, have proven valuable tools for analyzing the genetics of the CNS. One such mutant, called *spastic*, has a single locus recessive mutation on chromosome 3 and is characterized by strong behavioral deficits which develop about 2-3 weeks after birth: Tremor, hyperexcitability, and prolonged righting reflexes (Chai 1961). Genetic analysis has established the position of the *spastic* locus in linkage-group XVI of the mouse genome. Pharmacological, electromyographic, and biochemical data suggest that the *spastic* mutation affects GABA and glycine synapses in the entire CNS: (1) Subconvulsive doses of strychnine produce the *spastic* behavioral symptoms in normal mice; (2) ($^3$H)strychnine binding to *spastic* CNS membranes is reduced to 20%-30% of that of control littermates; and (3) the number of GABA/benzodiazepine receptors is elevated to about 140% of control animals (White and Heller 1982). This latter alteration is thought to reflect a compensatory response to the glycine-receptor deficiency.

In order to further characterize the underlying molecular defect, the pharmacology and biochemistry of the *spastic* glycine receptor has been characterized in more detail. In ($^3$H)strychnine-binding and agonist competition experiments, the *spastic* glycine receptor exhibited ligand affinities similar to the receptor protein in normal mice (White 1985, Becker et al. in preparation). Furthermore, after affinity purification the *spastic* receptor contained the three polypeptides of 48 K, 58 K, and 93 K characteristic of all mammalian glycine receptor preparations investigated so far (Becker et al. in preparation). Also, the immunological determinants of the glycine receptor defined by our mAbs were all conserved in the *spastic* mouse protein. These data strongly suggest that the *spastic* mutation defines a regulatory gene important for glycine-receptor expression and/or assembly/stabilization in the neuronal plasma membrane. Further characterization of the spastic mutation thus may help to elucidate the mechanisms by which the chemosensitivity of individual neurons is regulated in the mammalian CNS.

*Acknowledgments.* We thank C. Schröder and H. Krischke for expert technical assistence and Rosemary Franklin for help with the preparation of the manuscript. This work was supported by grants from the Deutsche Forschungsgemeinschaft, the Bundesministerium für Forschung und Technologie and the Fonds der Chemischen Industrie.

## References

Barker JL, McBurney RN (1979) GABA and glycine may share the same conductance channel on cultured mammalian neurones. Nature (London) 277:234-236

Betz H, Graham D, Pfeiffer F, Rehm H (1983) α-Bungarotoxin and strychnine as tools to characterize neurotransmitter receptors of the central nervous system. In: Hucho F, Ovchinnikov YA (eds) Toxins as tools in neurochemistry. De Gruyter, Berlin New York, pp 245-255

Chai CK (1961) Hereditary spasticity in mice. J Hered 52:241-243

Fritz LC, Wang CC, Gorio A (1979) Avermectin B1a irreversibly blocks postsynaptic potentials at the lobster neuromuscular junction by reducing muscle membrane resistance. Proc Natl Acad Sci USA 76:2062-2066

Graham D, Pfeiffer F, Betz H (1981) UV light-induced cross-linking of strychnine to the glycine receptor of rat spinal cord membranes. Biochem Biophys Res Commun 102:1330-1335

Graham D, Pfeiffer F, Betz H (1982) Avermectin B1a inhibits the binding of strychnine to the glycine receptor of rat spinal cord. Neurosci Lett 29:173-176

Graham D, Pfeiffer F, Betz H (1983) Photoaffinity-labelling of the glycine receptor of rat spinal cord. Eur J Biochem 131:519-525

Graham D, Pfeiffer F, Simler R, Betz H (1985) Purification and characterization of the glycine receptor of pig spinal cord. Biochemistry 24:990-994
Hamill OP, Bormann J, Sakmann B (1983) Activation of multiple-conductance state chloride channels in spinal neurones by glycine and GABA. Nature (London) 305: 805-808
Möhler H, Battersby MK, Richards JG (1980) Benzodiazepine receptor protein identified and visualized in brain tissue by a photoaffinity label. Proc Natl Acad Sci USA 77:1666-1670
Noda M, Takahashi H, Tanabe T, Toyosato M, Kikyotani S, Furutani Y, Hirose T, Takashima H, Inayama S, Miyata T, Numa S (1983) Structural homology of *Torpedo californica* acetylcholine receptor subunits. Nature (London) 302:528-532
Pfeiffer F, Betz H (1981) Solubilization of the glycine receptor from rat spinal cord. Brain Res 226:273-279
Pfeiffer F, Graham D, Betz H (1982) Purification by affinity chromatography of the glycine receptor of rat spinal cord. J Biol Chem 257:9389-9393
Pfeiffer F, Simler R, Grenningloh G, Betz H (1984) Monoclonal antibodies and peptide mapping reveal structural similarities between the subunits of the glycine receptor of rat spinal cord. Proc Natl Acad Sci USA 81:7224-7227
Sigel E, Barnard EA (1984) A γ-aminobutyric acid:benzodiazepine receptor complex from bovine cerebral cortex. J Biol Chem 259:7219-7223
Snyder SH, Bennett JP (1976) Neurotransmitter receptors in the brain: Biochemical identification. Annu Rev Physiol 38:153-175
St John PA, Owen DG, Barker JL, Pfeiffer F, Betz H (1984) Monoclonal antibodies to purified glycine receptor bind to glycine receptors of mouse spinal neurons in culture. Soc Neurosci Abstr vol 10, part I, p 6
Triller A, Cluzeaud F, Pfeiffer F, Korn H (1985) Localization of glycine receptors in rat spinal cord. J Cell Biol 101:683-688
White WF (1985) The glycine receptor in the mutant mouse spastic: strychnine binding characteristics and pharmacology. Brain Res 329:1-6
White WF, Heller AH (1982) Glycine receptor alterations in the mutant mouse spastic. Nature (London) 298:655-657
Young AB, Snyder SH (1973) Strychnine binding associated with glycine receptors of the central nervous system. Proc Natl Acad Sci USA 70:2832-2836
Young AB, Snyder SH (1974) The glycine synaptic receptor: Evidence that strychnine binding is associated with the ionic conductance mechanism. Proc Natl Acad Sci USA 71:4002-4005
Zarbin MA, Wamsley JK, Kuhar MJ (1981) Glycine receptor: Light microscopic autoradiographic localization with ($^3$H)strychnine. J Neurosci 1:532-547

# The GABA Receptor/Benzodiazepine Complex in the Central Nervous System

H. Möhler, P. Schoch, and J. G. Richards[1]

## Neuronal Inhibition in the Brain

One of the governing principles of neuronal communication in the central nervous system is the integration of neuronal excitation and inhibition. Various neurotransmitters, such as glutamate, aspartate, and certain neuropeptides, are able to stimulate neuronal activity. Inhibition of neuronal activity is exerted mainly by neurons operating with the neurotransmitter γ-aminobutyrate (GABA). Up to 30% of all synapses in the brain are thought to be GABAergic. They operate mainly in inhibitory feedback and feedforward circuits provided by GABAergic projecting neurons and local GABAergic interneurons (Roberts 1984). The synaptic inhibitory action of GABA is due to the opening of GABA-gated chloride channels leading to an increase in the chloride conductance of the subsynaptic membrane (Krnjevic 1976, Werman 1979, Peck 1980, Hamill et al. 1983, Simmonds 1984). Most frequently, the chloride flux is directed inward, leading to an inhibitory postsynaptic potential. Thereby, the excitability of the effector neuron is reduced.

In this article, we describe functional, structural, and histochemical aspects of GABA receptors associated with chloride channels. Such receptors are also termed $GABA_A$ receptors, to differentiate them from $GABA_B$ receptors which occur at low density in the brain, show an agonist and antagonist ligand specificity different from $GABA_A$ receptors and are not associated with chloride channels (Bowery et al. 1984). In the following, the term GABA receptor is synonymous with $GABA_A$ receptor.

## Regulation of GABA Receptors

The experimental investigation of the GABA receptor was greatly stimulated by the discovery that the receptor can be allosterically modulated by drugs acting on the so-called benzodiazepine receptor (BZR). This receptor was originally identified as the molecular target structure for benzodiazepines (Möhler and Okada 1977, Squires and Braestrup 1977), a group of drugs with wide therapeutic application as anxiolytics, hypnotics, muscle relaxants, and anticonvulsants; their main representative is diazepam (Valium). The BZR was found to be present in the central nervous system of all higher species (Möhler et al. 1978, Nielsen et al. 1978), where it serves as modulatory unit of the GABA receptor (for review Möhler 1984, Haefely et al. 1985). In the presence of tranquillizing benzodiazepines the efficiency of GABAergic synaptic transmission is enhanced (Haefely et al. 1978, Haefely 1984), result-

[1]Research Department, F. Hoffmann-La Roche & Co., Ltd., CH-4002 Basel, Switzerland

36. Colloquium - Mosbach 1985
Neurobiochemistry

ing in a shift to the left of the dose response curve for GABA (Choi et al. 1977, 1981, Haefely 1984). This modulatory action manifests itself as the main therapeutic effects of the tranquillizing benzodiazepines (Haefely and Polc 1985). All BZR ligands which, irrespective of their chemical structure, increase GABAergic synaptic transmission were termed BZR agonists. Such agonists act as positive allosteric effectors of the GABA receptor (Table 1).

Table 1. Modulation of CNS functions by BZR ligands

| Agonists | | Inverse agonists |
|---|---|---|
| - | Anxiety | + |
| - | Muscle tension | + |
| - | Convulsions | + |
| - | Vigilance | + |

In recent years, BZR ligands were found which did not enhance the action of GABA but resulted in a reduction of the efficiency of GABAergic synaptic transmission (Braestrup et al. 1982, Polc et al. 1982). These substances were termed inverse agonists of BZR; they act as negative allosteric effectors. In keeping with their negative efficacy on the synaptic level, the compounds show a spectrum of pharmacological effects which is opposite to that of BZR agonists. For instance, these agents are able to induce anxiety attacks, increase muscle tension, precipitate convulsions, and increase sleep latency (Table 1).

The effects of both types of BZR ligands, agonists and inverse agonists, are mediated via the same binding domain on the receptor. This could be most clearly demonstrated in competition experiments with the antagonist Ro 15-1788 (Hunkeler et al. 1981). The latter compound, despite a high affinity to the BZR, is largely devoid of pharmacological effects per se. But by competitive interaction it antagonizes the effects of both agonists and inverse agonists of the BZR (Möhler 1984, Haefely 1985).

These neurophysiological findings clearly show that opposite pharmacological effects are mediated via the same receptor site. Depending on the type of BZR ligand, the efficiency of GABAergic transmission can be altered in opposite directions, which has profound psychic and somatic consequences. Thus, the conformational state of the GABA receptor/BZR/chloride channel complex can determine the extent of anxiety, the degree of muscle tension, the likelihood fo convulsions, and the level of vigilance. To clarify the function of the GABA receptor/BZR/chloride channel complex on the molecular level, an attempt was made to determine its structure and cellular localization.

## Structural Analysis of the GABA Receptor/Benzodiazepine Receptor Complex

Our present knowledge on the structure and localization of the GABA receptor/BZR complex is based on radioligand binding experiments and studies with monoclonal antibodies (mAb). In binding studies, labeled benzodiazepines were particularly useful (for review Möhler and Richards 1983, Möhler 1984, Haefely et al. 1985). They (1) led to the discovery of BZR (Möhler and Okada 1977, Squires and Braestrup 1977), (2) allowed their autoradiographical visualization in brain tissue at the light- and electronmicroscopic level (Möhler et al. 1981, Richards and Möhler 1984, Young and Kuhar 1980), (3) demonstrated a receptor complex as shown by allosteric interactions between benzodiazepine binding sites, low affinity GABA sites, and binding sites for TBPS, a presumptive "chloride channel" ligand (for review Olsen et al. 1984, Supavilai and Karobath 1984, Haefely et al. 1985) and (4) allowed the identification of subunits of the benzodiazepine receptor by photo-affinity labeling (Möhler et al. 1980, Sieghart and Karobath 1980). Furthermore, benzodiazepine affinity resins permitted the purification of the GABA receptor complex (Sigel et al. 1983, Schoch and Möhler 1983, Schoch et al. 1984, Sigel and Barnard 1984, Kuriyama et al. 1984). Recently, subunit specific monoclonal antibodies (mAb) Häring et al. 1985, Schoch et al. 1985) allowed the identification of two types of subunits in the receptor (Häring et al. 1985) and their visualization in pre- and post-synaptic membranes in the central nervous system (Möhler et al. 1985, Richards et al. 1985).

In the following, the key developments leading to the present view on the structure and localization of the GABA receptor/BZR/chloride channel complex are described. The findings are summarized in a model of the receptor complex.

## Receptor Constituents Identified by Photoaffinity Labeling

The first information on the protein constituents of the GABA receptor/BZR complex was obtained by phtoreactive benzodiazepines. The benzodiazepine flunitrazepam, which is not light-sensitive per se, was found to be highly photoreactive when bound to the benzodiazepine receptor (Möhler et al. 1980). With this photoaffinity label a protein of 50 kd (α-subunit) was photolabeled in synaptic membrane preparations of all species and brain regions studied (Fig. 1) (Möhler et al. 1980). In a few brain regions, mainly in rat hippocampus, a second protein with a molecular weight of 55 kd (β-subunit) was photolabeled (Sieghart

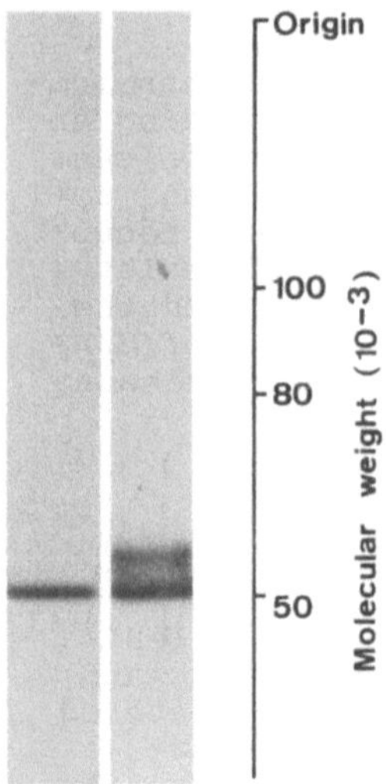


Fig. 1. Autoradiograph of photoaffinity labeled proteins separated by SDS-gel electrophoresis. *Left panel* rat cerebral cortex (Möhler et al. 1980); *right panel* rat hippocampus (Sieghart and Karobath 1980). Synaptic membranes were photolabeled with $^{3}$H-flunitrazepam

and Karobath 1980), suggesting an apparent structural heterogeneity of benzodiazepine receptors in different brain areas (Fig. 1).

The molecular association of GABA receptors with BZR was evident not only from electrophysiological studies but also on biochemical grounds. In synaptic membrane preparations, the affinity of BZR ligands was modified in the presence of micromolar concentrations of GABA. While the affinity of BZR agonists was enhanced, that of inverse agonists was reduced; the affinity of the competitive antagonist remained unaltered. Thus, the GABA-induced affinity shift seems to reflect the different biological efficacy of BZR ligands (for review Möhler and Richards 1983, Möhler 1984).

## Purification of the GABA Receptor Complex

Purification of the GABA receptor complex was expected to (1) yield more information on the constituents and structure of the receptor and (2) allow the production of monoclonal antibodies as additional tools for receptor analysis.

The receptor was solubilized by deoxycholate from membranes of bovine cerebral cortex and adsorbed to an affinity resin containing the benzodiazepine Ro 7-1986 coupled via a spacer to sepharose (Sigel et al. 1983). After specific elution with an excess of the water soluble benzodiazepine chlorazepate in the presence of Triton X-100/asolectin, the receptor was freed of chlorazepate by gel filtration and treatment with Biobeads. Although the isolation procedure is based on a benzodiazepine affinity resin, GABA binding sites were copurified. In radioligand binding assays, the approximately 1000-fold purified receptor preparation showed properties characteristic of the membrane-bound GABA receptor/BZR complex (Schoch et al. 1985) (Table 2). Benzodiazepine binding was stereospecific with a maximum number of $^3$H-flunitrazepam binding sites of 700 ± 120 pmol/mg protein and an apparent affinity constant $K_D = 2.5 \pm 0.5$ nM. The receptor fraction contained two types of binding sites for GABA: a high-affinity site with $K_D = 13.8 \pm 8.0$ nM for $^3$H-muscimol (maximum number 1020 ± 310 pmol/mg protein) and a low-affinity site which was monitored by the modulation of benzodiazepine receptor affinity induced by high concentrations of GABA. In the presence of 1 μM GABA, the affinity of the agonist $^3$H-diazepam was enhanced while that of the inverse agonist $^3$H-βCCM was reduced as compared to controls to which no GABA was added. In contrast, the affinity of the competitive antagonist $^3$H-Ro 15-1788 remained unaltered by GABA (Table 3).

Thus, three types of binding sites which are associated with the membrane-bound GABA receptor complex were present in the purified receptor preparation. The presence of the low affinity GABA sites was especially pertinent since such sites are of functional significance in gating the chloride channel (Werman 1979, Sakmann et al. 1983).

In SDS gel electrophoresis, the receptor preparation showed Coomassie Blue-stained bands of 50 kd, 55 kd, and a more diffuse band of 62 kd. The 50 kd band could be photoaffinity labeled with $^3$H-flunitrazepam, indicating that these proteins were constituents of the GABA receptor/BZR complex (Schoch et al. 1984). Using this purified receptor preparation attempts were made to reconstitute the receptor complex in lipid bilayers and to produce mAB directed against the receptor constituents.

Table 2. Binding characteristics of the purified GABA receptor/benzodiazepine receptor preparation

| A | $K_D$, $K_i$ (nM) | | $B_{max}$ (pmol/mg protein) |
|---|---|---|---|
| $^3$H-Flunitrazepam | 2.5 ± 0.5 | | 700 ± 120 |
| Diazepam | 15 | | |
| Benzodiazepine enantiomers | | | |
| Ro 11-6896(+)(active) | 3.5 | | |
| Ro 11-6893(-)(inactive) | > 1000 | | |
| Ro 15-1788 | 0.4 | | |
| β-CCE | 0.9 | | |
| β-CCM | 4.5 | | |
| Ro 5-4864 | > 1000 | | |
| $^3$H-Muscimol | 13.8 ± 8.0 | | 1020 ± 310 |
| GABA | 45 | | |
| Baclofen | > 1000 | | |
| B | $K_D$ (nM) | | |
| | - GABA | + GABA | |
| $^3$H-Diazepam | 24 | 12 | |
| $^3$H-Ro 15-1788 | 1.4 | 1.3 | |
| $^3$H-β-CCM | 2.4 | 5.2 | |

A Affinity ($K_D$) and inhibition ($K_i$) constants of various compounds for the benzodiazepine binding site ($^3$H-flunitrazepam binding) and the high affinity GABA site ($^3$H-muscimol binding).
B Effect of GABA (1 μM) on the apparent affinity constant of an agonist ($^3$H-diazepam), an antagonist ($^3$H-Ro 15-1788) and an inverse agonist ($^3$H-β-CCM) of the benzodiazepine receptor. The maximal number of binding sites ($B_{max}$) did not change in the presence of GABA. For details see Schoch and Möhler (1983) and Schoch et al. (1984)

## Reconstitution

In addition to the binding site for GABA and benzodiazepines, the native GABA receptor/benzodiazepine receptor complex contains a chloride ionophore. Binding of $^{35}$S-TBPS (t-butyl-bicyclophosphorothionate) is thought to be a biochemical marker of the ionophore, since there is evidence that this compound blocks a gate for the chloride channel (Squires et al. 1983, Supavilai and Karobath 1984). However, a functional analysis of the ionophore requires the reconstitution of the

Table 3. Reconstitution of the purified GABA-benzodiazepine receptor in asolectin vesicles

| Preparation | $^3$H-Flunitrazepam binding activity |
|---|---|
| Receptor sample before reconstitution | 100% (415 pmole/mg protein) |
| Reconstituted receptor | 54% |
| Reconstituted receptor after sedimentation | 35% |

Reconstitution of the purified receptor into asolectin vesicles was accomplished by the Bio-Beads technique. $^3$H-Flunitrazepam binding was determined in the reconstituted fraction and in the pellet after centrifugation at 270,000 *g* for 90 min (Schoch et al. 1984)

purified receptor complex in lipid bilayers. By adding Bio-Beads SM-2 to a bovine receptor preparation which contained Triton X-100 and soybean lecithin, the benzodiazepine receptor could be reconstituted in lipid vesicles (Schoch et al. 1984). Due to the adsorption of Triton X-100 to the Bio-Beads, small lipid vesicles were formed which contained 54% of the $^3$H-flunitrazepam binding activity. About two thirds of this activity sedimented with the vesicle fraction at 270,000 *g*, giving a total of 35% of the initial binding sites that are reconstituted. Thus, the benzodiazepine receptor was indeed incorporated into lipid bilayers in a functional form (Schoch et al. 1984). Receptor reconstitution is a prerequisite to study the structural requirements for chloride channel function.

## Analysis of the Receptor Complex with Subunit-Specific Monoclonal Antibodies

Recently, a collection of 16 mAb directed against the GABA receptor complex was produced (Häring et al. 1985, Schoch et al. 1985). The specificity of the mAb is based on five criteria:

1. In immunoprecipitation experiments (Häring et al. 1985) a protein oligomer containing the binding sites for benzodiazepines, GABA (high and low affinity) (Table 4), and TBPS (Schoch, in preparation) was precipitated.
2. In rat brain sections the distribution and density of the immunostain corresponded almost entirely to the distribution of binding sites for benzodiazepines (Fig. 2). Areas which lacked benzodiazepine receptors, such as white matter, pineal gland, or pituitary, were also devoid of immunoreactivity (Schoch et al. 1985).
3. Peripheral organs which contain no benzodiazepine receptors, such as adrenal, kidney, or superior cervical ganglion, were devoid of immunoreactivity (Schoch et al. 1985).
4. Immunoreactive polypeptides, identified in immuno-blots of bovine, rat and human receptor preparations, showed molecular weights of 50 kd and 55 kd (Fig. 3). These values correspond to those of the known subunits of the GABA receptor/BZR complex. The mAb were subunit specific (Fig. 3) (Häring et al. 1985).

Table 4. Immunoprecipitation of radioligand binding sites of the GABA receptor/benzodiazepine receptor complex

| Epitope | Monoclonal antibody | Radioligand recovered in the pellet (%) $^3$H-Flunitrazepam binding[a] | $^3$H-Muscimol binding |
|---|---|---|---|
| I | bd-17 | 91 | 92 |
| II | bd-24 | 98 | 91 |
| III | bd-8 | 95 | 96 |
| IV | bd-28[b] | 49 | 41 |
| | Unrelated mAb | 7 | 6 |

Immunoprecipitation was performed (Häring et al. 1985) by incubating a purified receptor fraction of bovine cerebral cortex with purified mAb bd-17 and bd-28 or with hybridoma supernatant (bd-8, bd-28) followed by goat anti-mouse immunosorbent. The immunoprecipitated and nonprecipitated receptor fractions were assayed by $^3$H-flunitrazepam (10 nM, 60 min) and $^3$H-muscimol (100 nM, 30 min) binding at 4°C. Controls were performed with an irrelevant antibody. Values are means of two experiments. The mAb given are representative for four different groups of mAb recognizing four different epitopes (I-IV) of the receptor complex. Immunoprecipitation of the GABA receptor/BZR complex from rat whole brain yielded similar results except for group II mAb which showed only background levels (see Table 5). Receptor preparations from human cerebral cortex could be immunoprecipitated to a similar extent as bovine receptor (Häring et al. 1985).

[a] In the presence of 10 μM GABA $^3$H-flunitrazepam binding was enhanced in all cases by 20%-40% indicating the presence of low affinity GABA sites in the pellet (Häring et al. 1985).

[b] The immunoprecipitation was incomplete due to nonsaturating concentration of bd-28 in the hybridoma supernatant (Häring et al. 1985).

5. The mAb could be divided into four groups recognizing four different epitopes of the receptor complex as shown most clearly by a comparison of their subunit-specificity and their species specificity (Table 5) (Häring et al. 1985, Schoch et al. 1985).

The following biochemical and morphological properties of the GABA receptor complex could be established using the subunit specific mAb (Häring et al. 1985, Möhler et al. 1985, Schoch et al. 1985):

1. The binding sites for benzodiazepines, GABA (high and low affinity), and TBPS are localized on the same protein complex as shown by immunoprecipitation (Häring et al. 1985, Schoch et al. 1985).
2. Each receptor complex contains both α- and β-subunits as shown by immunoprecipitation of bovine receptor preparations. Receptor complexes consisting exclusively of α-subunits or β-subunits do not seem to exist (Häring et al. 1985, Möhler et al. 1985).
3. Topographically, both α- and β-subunits seem to be exposed on the outer surface of the cell as indicated by electron microscopic immunocytochemistry in rat brain (Richards, in preparation).
4. Most, if not all, BZR appear to be part of GABA receptors. This was demonstrated in rat brain by the extensive co-localization of benzo-

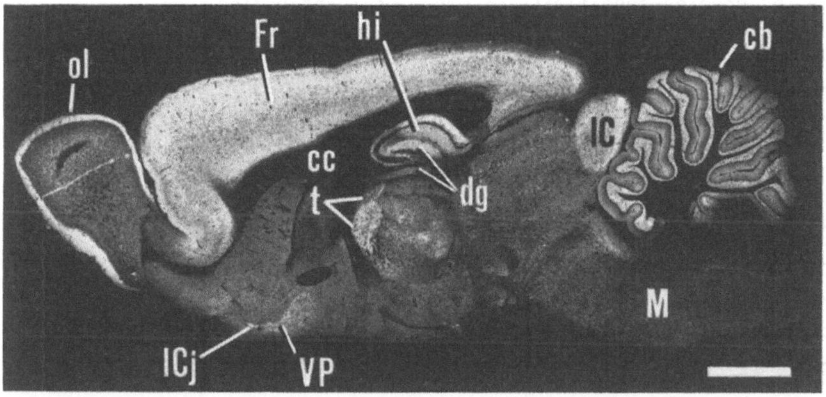


Fig. 2. Immunohistochemical localization of antigenic sites in a parasaggital section of rat brain using the mAb bd-17 (Schoch et al. 1985). The staining pattern is very similar to the density and distribution of benzodiazepine receptors visualized autoradiographically by radioligand binding. In the granular layer of the cerebellum and in the ventro-lateral thalamus the immune reaction was more intense than the autoradiographical pattern of benzodiazepine binding sites, but corresponded to that of high-affinity GABA sites and TBPS binding sites (Schoch et al. 1985). The receptors in those two areas may occur in a conformation with a predominance of high-affinity GABA sites and low-affinity benzodiazepine binding sites. Bar = 2 mm. *cb* cerebellum; *cc* corpus callosum; *Fr* frontal cerebral cortex; *dg* dentate gyrus; *hi* hippocampus; *ICj* islands of Calleja; *IC* inferior colliculus; *M* medulla; *ol* olfactory bulb; *t* thalamus; *VP* ventral pallidum

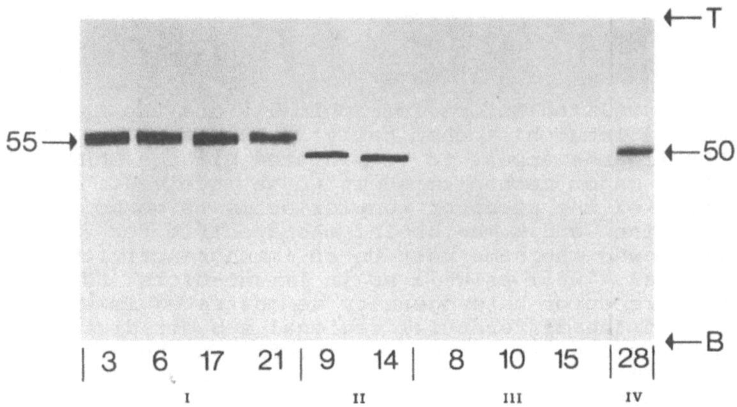


Fig. 3. Autoradiograph of an immuno-blot with the GABA receptor/benzodiazepine receptor complex purified from bovine cerebral cortex (Häring et al. 1985). mAb representing all four epitopes (*I–IV*) are indicated by the number (bd-x) at the bottom of each strip. Top (*T*), bottom (*B*) of the gel. The position of 50 kd and 55 kd proteins is indicated (Häring et al. 1985). Although the mAb of epitope III were reactive in immunoprecipitation (Table 4) and in solid phase antibody binding assays, their subunit specificity could not be determined since this epitope was sensitive to the denaturing conditions of immuno-blots (Häring et al. 1985)

Table 5. Subunit- and species-specificity of monoclonal antibodies directed against the GABA receptor/benzodiazepine receptor complex

| Epitope | Monoclonal antibody | Subunit specificity | |
|---|---|---|---|
| | | Human/bovine receptor | Rat receptor |
| I | bd-17 | β | β |
| II | bd-24 | α | n.i.[a] |
| III | bd-8 | n.d.[b] | n.d.[b] |
| IV | bd-28 | α | α |

For each epitope (I-IV) one representative mAb is given. The subunit specificity was determined in immuno-blots using receptor preparations from human cerebral cortex, bovine cerebral cortex and cerebellum, and rat cerebral cortex, hippocampus and cerebellum. (Häring et al. 1985).

[a]Group II mAb did not react with rat receptor in either solid phase antibody binding assays, immunoprecipitation, immunoblotting, and immunohistochemistry (n.i. = no immune response)(Häring et al. 1985, Schoch et al. 1985).

[b]Although the mAb of epitope III were reactive in immunoprecipitation and solid phase antibody binding assays, their subunit specificity could not be determined (n.d.) since the epitope was sensitive to the denaturing conditions in immunoblots (Häring et al. 1985).

diazepine binding sites visualized autoradiographically and the GABA receptor complex visualized immunohistochemically (Schoch et al. 1985). Thus, the effects of BZR ligands appear to be mediated via the GABA receptor complex; no other major mechanism seems to be involved.
5. The subunit composition of the receptor complex seems to be homogenous throughout rat, bovine, and human brain. mAb specific for either the α- or the β-subunit showed the same pattern of immunoreactivity in brain sections (Schoch et al. 1985) as well as in immuno-blots (Häring et al. 1985). The lack of receptor heterogeneity demonstrated immunologically is in contrast to the differential regional subunit distribution obtained by photolabeling (see above). It remains to be clarified why photolabeling of the β-subunit occurs in rat hippocampus but not in other parts of the brain.
6. In keeping with the presence of postsynaptic GABA receptor complexes in the brain, immunoreactivity could be visualized on postsynaptic membranes by electron microscopic immunocytochemistry (Fig. 4) (Möhler et al. 1985, Richards et al. 1985).
7. GABA receptor complexes seem to occur also on GABAergic nerve endings. In those synapses in which the postsynaptic membrane was stained, the presynaptic membrane also showed immunoreactivity (Fig. 4) (Möhler et al. 1985). This finding sheds new light on the mechanism of action of BZR ligands. They appear to act not only on postsynaptic GABA receptors but also on GABA autoreceptors.

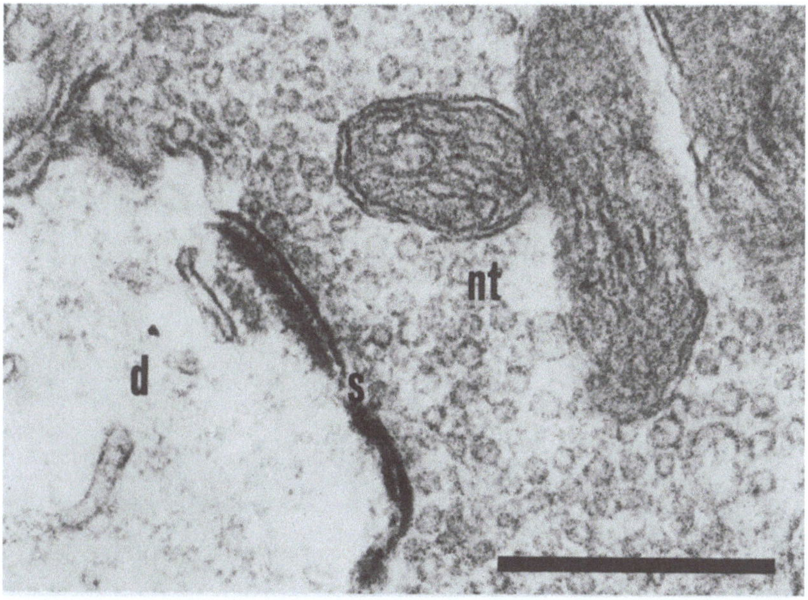


Fig. 4. Electronmicroscopic localization of antigenic sites in rat substantia nigra zona reticulata. The electron dense peroxidase immune reaction is present exclusively in the pre- and postsynaptic membrane, as shown for an axodendritic contact. *nt* nerve terminal; *s* synapse; *d* dendrite. Bar = 0.5 μm

## Model of the GABA Receptor Complex

Although many questions about the molecular structure of the GABA receptor complex are still unanswered, the known features can be incorporated into a preliminary working hypothesis (Fig. 5). There is unequivocal immunological evidence that the receptor contains both α- and β-subunits. A tetrameric $\alpha_2/\beta_2$ subunit arrangement is suggested by the comparison of the molecular weight of the subunits and the native receptor. While the α- and β-subunits have a molecular weight of 50 kd and 55 kd, respectively (Möhler et al. 1980, Sieghart and Karobath 1980), the target size for benzodiazepine and GABA binding sites in membranes was 220 kd determined by radiation inactivation (Chang and Barnard 1982). Furthermore, a tetrameric model was also indicated by photoaffinity labeling experiments in which photoreaction to one benzodiazepine binding site in the membrane caused a conformational affinity change in three additional sites (Möhler et al. 1980). An $\alpha_2/\beta_2$ structure might also accomodate the TBPS binding site as shown by immunoprecipitation (see above), and by radioligand binding in a receptor preparation purified in CHAPS detergent (Sigel and Barnard 1984). It is thus conceivable that the $\alpha_2/\beta_2$ tetramer contains not only all the known binding sites of the receptor complex but also forms the chloride channel (Fig. 5). A protrusion of the protein on the outer cell surface is suggested by the susceptibility of benzodiazepine binding sites to the action of proteases (Klotz et al. 1984). The receptor is a glycoprotein (Gavish and Snyder 1981, Möhler, unpublished).

There is only limited information which allows on allocation of the various ligand binding sites to certain receptor domains. Binding sites

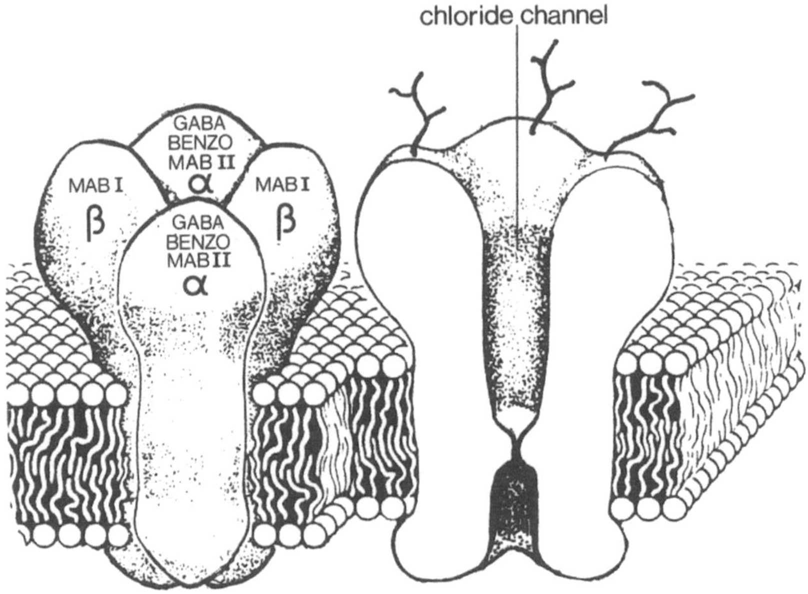


Fig. 5. Working hypothesis for a model of the GABA receptor/benzodiazepine receptor/chloride channel-complex (Möhler et al. 1985). For explanation see text. α and β indicate the 50 kg and 55 kd subunits, respectively. *Benzo* benzodiazepine binding sites identified by photolabeling. The presence of benzodiazepine binding sites on the β-subunit (generally not photolabeled) cannot be excluded. *GABA* GABA binding sites identified by photolabeling. Nonphotolabeled GABA sites on the β-subunit cannot be excluded. *MAB I, II* indicate subunit-specific epitopes recognized by group I and group II mAb, respectively, on the outer receptor domain. *Branched lines* carbohydrate residues

for benzodiazepines are clearly located on the α-subunit (Möhler et al. 1980). Conceivably, they might also occur on the β-subunit, although in a state which is generally not photolabeled (Sieghart and Karobath 1980). Electrophysiologically, the presence of two to four GABA sites was postulated per GABA receptor to gate chloride channels (Hamill et al. 1983, Sakmann et al. 1983, Werman 1979). Biochemically, GABA sites were attributed to the α-subunit by $^3$H-muscimol photolabeling (Asano et al. 1983, Cavalla and Neff 1985), although the possible presence of non-photolabeled GABA sites on the β-subunit cannot be excluded. The location of the barbiturate and TBPS binding sites is yet unknown. Binding sites may well occur in subunit interfaces.

Some experimental results suggest that the present working hypothesis on the receptor structure (Fig. 5) might have to be modified in the future. For instance, an additional protein of 62 kd occurs in our purified receptor preparations (Schoch et al. 1984). Furthermore, the target size of the radiation-inactivated TBPS binding site was found to be 134 kd (Nielsen and Braestrup 1983). Both results might be explained by the presence of an additional subunit of 62-80 kd (γ-subunit in the receptor complex.

## Outlook

1. New information on the synthesis, assembly, and structure of the receptor is expected to arise in the near future from the isolation and analysis of the GABA receptor genes. Such studies will also provide DNA probes to identify by in situ hybridization those cells which express the GABA receptor genes and are thus the target cells for GABAergic regulation.
2. The diagnostics of possible CNS disease which may be associated with dysfunction of the GABA receptor complex were so far restricted to BZR radioligand autoradiography of post mortem tissue. Now, possible alteration of the receptor can be monitored by immunohistochemistry in human brain, a technique which is not influenced by the possible presence of benzodiazepines in the tissue or altered GABA levels.
3. The physiological role of the BZR is yet unknown. The possible presence in the brain of endogenous ligands for this receptor remains an attractive hypothesis (Alho et al. 1985).

*Acknowledgment.* We would like to thank Dr. W. Haefely for critical reading of the manuscript.

## References

Alho H, Costa E, Ferrero P, Fujimoto M, Cosenza-Murphy D, Guidotti A (1985) Diazepam binding inhibitor (DBI): a neuropeptide located in selected neuronal populations of rat brain. Science 229:179-182

Asano T, Yamada Y, Ogasawara N (1983) Characterization of the solubilized GABA and benzodiazepine receptor from various regions of bovine brain. J Neurochem 40: 209-214

Bowery NG, Hill DR, Hudson AL, Price GW, Turnbull MJ, Wilkin GP (1984) Heterogeneity of mammalian GABA receptors. In: Bowery NG (ed) Actions and interactions of GABA and benzodiazepines. Raven Press, New York, pp 81-108

Braestrup C, Schmiechen, R, Neff G, Nielsen M, Petersen EN (1982) Interaction of convulsive ligands with benzodiazepine receptors. Science 216:1241-1243

Cavalla D, Neff NH (1985) Photoaffinity labeling of the $GABA_A$ receptor with $^3$H-muscimol. J Neurochem 44:916-921

Chang L-R, Barnard EA (1982) The benzodiazepine/GABA receptor complex: molecular size in brain synaptic membranes and in solution. J Neurochem 39:1507-1518

Choi DW, Farb DH, Fischbach GD (1977) Chlordiazepoxide selectively augments GABA action in spinal cord cell cultures. Nature (London) 269:342-344

Choi DW, Farb DH, Fischbach GD (1981) Chlordiazepoxide selectively potentiates GABA conductance of spinal cord and sensory neurons in cell culture. J Neurophysiol 45:621-631

Gavish M, Snyder SH (1981) γ-Aminobutyric acid and benzodiazepine receptors: copurification and characterization. Proc Natl Acad Sci USA 78:1939-1942

Haefely W (1984) Actions and interactions of benzodiazepine agonists and antagonists at GABergic synapses. In: Bowery NG (ed) Actions and interactions of GABA and benzodiazepines. Raven Press, New York, pp 263-285

Haefely W (1985) Pharmacology of benzodiazepine antagonists. Pharmacopsychiatry 18:163-166

Haefely W, Polc P (1985) Electrophysiological studies on the interactions of anxiolytic drugs with GABAergic mechanisms. In: Olsen RW, Venter JC (eds) Benzodiazepine — GABA receptors and chloride channels: Structural and functional properties. Liss, New York, in press

Haefely W, Polc P, Schaffner R, Keller HH, Pieri L, Möhler H (1978) Facilitation of GABAergic transmission by drugs. In: Krogsgaard-Larsen P, Scheel-Krüger J, Kofod H (eds) GABA neurotransmitters. Munksgaard, Copenhagen, pp 357-375

Haefely W, Kyburz E, Gerecke M, Möhler H (1985) Recent advances in the molecular pharmacology of benzodiazepine receptors and in the structure activity relationships of their agonists and antagonists. In: Advances in drug research. In press
Häring P, Stähli C, Schoch P, Takács B, Staehelin T, Möhler H (1985) Monoclonal antibodies reveal structural homogeneity of γ-aminobutyric acid/benzodiazepine receptors in different brain areas. Proc Natl Acad Sci USA 82:4837-4841
Hamill OP, Bormann J, Sakmann B (1983) Activation of multiple conductance state chloride channels in spinal neurons by glycine and GABA. Nature (London) 305: 805-808
Hunkeler W, Möhler H, Pieri L, Polc P, Bonetti EP, Cumin R, Schaffner R, Haefely W (1981) Selective antagonists of benzodiazepines. Nature (London) 290:514-516
Klotz KL, Bocchetta A, Neale JH, Thomas JW, Tallman JF (1984) Proteolytic degradation of neuronal benzodiazepine binding sites. Life Sci 34:293-299
Krnjevic K (1976) Inhibitory action of GABA and GABA-mimetics on vertebrate neurons. In: Roberts E (ed) GABA in nervous system function. Kroc Found Ser, vol V. Raven Press, New York, pp 269-282
Kuriyama K, Youeda Y, Taguchi J, Takahashi M, Ohkuma S (1984) Properties of purified γ-aminobutyric (GABA) receptor and modulation of GABA receptor binding by membrane phospholipids. Neuropharmacology 23:839-840
Möhler H (1984) Benzodiazepine receptors and their ligands. In: Bowery NG (ed) Actions and interactions of GABA and benzodiazepines. Raven Press, New York, pp 155-166
Möhler H, Okada T (1977) Demonstration of benzodiazepine receptors in the central vervous system. Science 198:849-851
Möhler H, Richards JG (1983) Receptors for anxiolytic drugs. In: Malick JB, Enna SJ, Yamamura HI (eds) Anxiolytics, neurochemical, behavioural and clinical perspectives. Raven Press, New York, pp 15-40
Möhler H, Okada T, Heitz P, Ulrich J (1978) Biochemical identification of the site of action of benzodiazepines in human brain by $^{3}$H-diazepam binding. Life Sci 22: 985-996
Möhler H, Battersby MK, Richards JG (1980) Benzodiazepine receptor protein identified and visualized in brain tissue by a photoaffinity label. Proc Natl Acad Sci USA 77:1666-1670
Möhler H, Richards JG, Wu J-Y (1981) Autoradiographic localization of benzodiazepine receptors in immunocytochemically identified γ-aminobutyrergic synapses. Proc Natl Acad Sci USA 78:1935-1938
Möhler H, Schoch P, Richards JG, Häring P, Takács B, Stähli C (1985) Monoclonal antibodies as probes for studying the structure and location of the GABA receptor/ benzodiazepine receptor/chloride channel complex. In: Olsen RW, Venter JC (eds) Benzodiazepine – GABA receptors and chloride channels: Structural and functional properties. Liss, New York, in press
Nielsen M, Braestrup C (1983) The molecular target size of brain TBPS binding sites. Eur J Pharmacol 91:321-322
Nielsen M, Braestrup C, Squires RF (1978) Evidence for a late evolutionary appearance of brain-specific benzodiazepine receptors: an investigation of 18 vertebrate and 5 invertebrate species. Brain Res 141:342-346
Olsen RW, Wong EHF, Stauber GB, Murakami D, King RG, Fischer JB (1984) Biochemical properties of the GABA/barbiturate/benzodiazepine receptor-chloride ion channel complex. In: Kito S, Segawa T, Kuriyama K, Yamamura HI, Olsen RW (eds) Neurotransmitter receptors. Plenum Press, New York, pp 205-219
Peck EJ (1980) Receptors for amino acids. Annu Rev Physiol 42:615-627
Polc P, Bonetti EP, Schaffner R, Haefely W (1982) A three state model of the benzodiazepine receptor explains the interactions between the benzodiazepine antagonist Ro 15-1788, benzodiazepine tranquilizers, β-carboline and phenobarbitone. Naunyn Schmiedeberg's Arch Pharmacol 321:260-264
Richards JG, Möhler H (1984) Benzodiazepine receptors. Neuropharmacology 23:233-242
Richards JG, Schoch P, Möhler H, Haefely W (1986) Benzodiazepine receptors resolved. Experientia, in press
Roberts E (1984) γ-Aminobutyric acid (GABA): from discovery to visualization of GABAergic neurons in the vertebrate nervous system. In: Bowery NG (ed) Actions and interactions of GABA and benzodiazepines. Raven Press, New York, pp 1-25

Sakmann B, Hamill OP, Bormann J (1983) Patch-clamp measurements of elementary chloride currents activated by putative inhibitory transmitters GABA and glycine in mammalian spinal neurons. J Neural Transm Suppl 18:83-95
Schoch P, Möhler H (1983) Purified benzodiazepine receptor retains modulation by GABA. Eur J Pharmacol 95:323-324
Schoch P, Häring P, Takács B, Stähli C, Möhler H (1984) A GABA/benzodiazepine receptor from bovine brain: purification, reconstitution and immunological characterization. J Recept Res 4:189-200
Schoch P, Richards JG, Häring P, Takács B, Stähli C, Staehelin T, Haefely W, Möhler H (1985) Co-localization of $GABA_A$ receptors and benzodiazepine receptors in the brain shown by monoclonal antibodies. Nature (London) 314:168-170
Sieghart W, Karobath M (1980) Molecular heterogeneity of benzodiazepine receptors. Nature (London) 286:285-287
Sigel E, Barnard EA (1984) A γ-aminobutyric acid/benzodiazepine receptor complex of bovine cerebral cortex: Improved purification with preservation of regulatory sites and their interactions. J Biol Chem 259:7219-7223
Sigel E, Stephenson A, Mamalaki C, Barnard EA (1983) A γ-aminobutyric acid/benzodiazepine receptor complex of bovine cerebral cortex: Purification and partial characterization. J Biol Chem 258:6965-6971
Simmonds MA (1984) Physiological and pharmacological characterization of the actions of GABA. In: Bowery NG (ed) Actions and interactions of GABA and benzodiazepines. Raven Press, New York, pp 27-41
Squires RF, Braestrup C (1977) Benzodiazepine receptors in rat brain. Nature (London) 266:732-734
Squires RF, Casida JE, Richardson M, Saedrup E (1983) [$^{35}$S]t-Butylbicyclophosphorothionate binds with high affinity to brain-specific sites coupled to γ-aminobutyric acid-A and ion recognition sites. Mol Pharmacol 23:326-336
Supavilai P, Karobath M (1984) [$^{35}$S]t-Butylbicyclophosphorothionate binding sites are constituents of the γ-aminobutyric acid benzodiazepine receptor complex. J Neurosci 4:1193-1200
Werman R (1979) Stoichiometry of GABA-receptor interactions: GABA modulates the glycine-receptor interaction allosterically in a vertebrate neuron. In: Mandel P, de Feudis FV (eds) GABA-biochemistry and CNS function. Plenum Press, New York, pp 287-301
Young WS III, Kuhar MJ (1980) Radiohistochemical localization of benzodiazepine receptors in rat brain. J Pharmacol Exp Ther 212:337-346

# Myasthenia Gravis: State of the Art and Approaches to Immunotherapy

T. Barkas[1], J.-M. Gabriel[1], M. Juillerat[1], M. Ballivet[1], and S. J. Tzartos[2]

## Myasthenia Gravis

### Clinical Features

Myasthenia gravis is a neuromuscular disorder characterized by fluctuating weakness and easy fatigability of voluntary muscles. Characteristically, the patient first presents with weakness of the ocular muscles (60%, drooping of eyelids, double vision), of the bulbar muscles (20%, difficulty in speech and swallowing), or of the limbs (20%, e.g., inability to keep arm outstretched). The signs may remain localized, but generally spread to include other muscle groups. Three types of myasthenia gravis are recognized, these being "acquired" (the most common and developing after birth), "neonatal" (a short-lasting myasthenia occurring in babies born to myasthenic mothers), and "congenital" (long-lasting and occurring in babies born to nonmyasthenic mothers). The prevalence is variously reported as 12-64 per million of the population, and the annual incidence rate as 2-4 per million (reviewed by Oosterhuis 1984). Adult patients fall into two categories, one of age 20-40 years, predominantly female, associated with HLA type B8 in Caucasians and B12 in Japanese, and DR3, and a second group of age 40-60 years with an equal sex distribution, no association with a specific HLA type, but characterized by the presence of thymomas (see Oosterhuis 1984).

### Historical Perspective – Causes of Disease and Treatment

A disease which can be interpreted as myasthenia gravis was first described in 1672 by Willis (Willis 1672). "Modern" myasthenia gravis dates from the description by Erb (Erb 1879) and Goldflam (Goldflam 1893) in 1879 and 1893, respectively, of small groups of patients suffering from fluctuating muscle weakness with fatigability of the muscle. The latter observation was recorded electrophysiologically by Jolly (Jolly 1895) in 1895. In 1901, Weigert (Weigert 1901) described a patient with a thymoma and cellular infiltration of muscle, suggesting a relationship between the thymus and the disease. In 1912, Sauerbruch (Schumacher and Roth 1912) performed thymectomy on a myasthenic patient with resulting improvement of the myasthenia; this approach was continued by Blalock in the period 1939-1941 (Blalock et al. 1939, Blalock 1944). Localization of the disease to the neuromuscular junction was first suggested by Walker in 1934, who demonstrated that anticholinesterase agents could be effective as a therapy (Walker 1934, 1935). Moving to 1960, an autoimmune origin for myasthenia gravis was suggested by Simpson (Simpson 1960) and Nastuk, the former on the basis

---

[1]Département de Biochimie, Université de Genève, Geneva, Switzerland
[2]Institut Pasteur Hellenique, Athens, Greece

36. Colloquium - Mosbach 1985
Neurobiochemistry

of clinical findings and the latter on the observations that some sera from myasthenic patients were toxic for muscle (Nastuk et al. 1959) and that complement levels fluctuated in relation to the disease state (Nastuk et al. 1960). Immunosuppressive agents were brought into use in 1965-1970 (von Reis et al. 1963, Delwaide et al. 1967, Warmolts et al. 1970). A postsynaptic defect was suggested in 1971-1973 by Engel (Engel and Santa 1971) and Fambrough (Fambrough et al. 1973) and this was firmly established in 1973 by the observation by Patrick and Lindstrom that immunization of rabbits with purified nicotinic acetylcholine receptor (nAChR) produced an experimental disease with many similarities to myasthenia (Patrick and Lindstrom 1973). This led rapidly to the demonstration of antibodies to the human nAChR in the sera of myasthenic patients (Lindstrom et al. 1976c) and to the development of plasma exchange in combination with immunosuppressive drugs as a means of therapy (e.g., Dau 1981).

## Experimental Autoimmune Myasthenia Gravis (EAMG)

EAMG can be induced in a range of species by immunization with homologous or heterologous nAChR, thus providing convenient models for studies on myasthenia gravis and possible immunotherapy. EAMG is also of wider interest in immunology, as it represents a useful model for studies of autoimmunity to a highly characterized antigen and immune regulation in general. The species generally used are the rabbit, the rat, and the mouse.

The rat model has been most studied. Young female Lewis rats are injected intradermally at multiple sites with purified nAChR (normally from the electric ray, *Torpedo*) in Freund's complete adjuvant containing *Mycobacterium tuberculosis* and *M. butyricum*, and, as a further adjuvant, *Bordetella pertussis* is given subcutaneously (Lennon et al. 1975, 1976). When large quantities (100 μg) of nAChR are given, an acute phase of muscle weakness can be observed after about seven days. The animals then recover, but start to develop a chronic form of weakness four weeks after immunization. Weakness is predominantly of the forelimbs, leading to hunching of the back and feeble grip of the forepaws, spreading to general paralysis. When smaller amounts (10-20 μg) of antigen are used, the initial acute phase is absent. Anti-nAChR antibodies peak at about four weeks after immunization; in the acute phase, their levels are low (Lindstrom et al. 1976b). At the microscopic level, the acute phase is characterized by infiltration of the synapse by phagocytic cells and destruction of the postsynaptic membrane (Engel et al. 1976). This phase has been shown to be complement-dependent (Lennon et al. 1978) and the components of complement can be seen on the membrane (Engel et al. 1979). Taken together with the observation of terminal components of complement on the membrane in myasthenic patients (Sahashi et al. 1980), this is often interpreted as demonstrating that at least a proportion of the damage is directly mediated by complement attack of the membrane. However, the complement dependency of the acute phase might simply be explained by IgM activation of the complement cascade as far as C3, resulting in attraction of phagocytic cells (C3a) and their stimulation due to opsonization (C3b). In this case, the observed binding of the terminal components might be incidental. Such an interpretation would be in agreement with results in the mouse model in which passive transfer of disease using myasthenic IgG was possible in the absence of C5 but not in the absence of C3 (Toyka et al. 1977).

In the rabbit and mouse models, animals are normally immunized twice with 100 μg (rabbit) or 10 μg (mouse) of nAChR. The rabbit model is useful in that the symptoms are highly reproducible in onset. The protocol used in our laboratory is to immunize intramuscularly with 100 μg of receptor in Freund's complete adjuvant (Gibco) and to boost three weeks later with the same amount of antigen in incomplete adjuvant. Six to eight days later, weakness of the hind limbs is observed. These are drawn forward resulting in a hunched appearence of the resting rabbit. Within 24 hours, this progresses through paralysis of the hindlimbs to complete paralysis of the animal as demonstrated by an inability of the rabbit to right itself if placed on its back. The mouse model has been used extensively to show a dependency of EAMG on the H2 type, in particular the I-A region (Berman and Patrick 1980a,b, Christadoss et al. 1979, 1982).

The model chosen for a particular purpose will obviously depend on the object of the experiment and the means available. In the rabbit model, visible signs of weakness are highly reproducible and obvious, whereas, in the rat and mouse models, this is not the case, due to the enhanced release of acetylcholine per stimulus in these species (it is not uncommon to observe 40%-60% reduction of receptor levels in these species without the presence of visible signs of weakness). At least in our hands, testing the animals for fatigability by exercise is not reliable. To quantitate results in these species, it may be necessary to carry out electrophysiological testing or to sacrifice animals and measure total body content of receptor or sensitivity to curare. However, these two species do have the obvious advantages that larger numbers of animals can be used to test the significance of a given result, that classical genetics and immunology experiments can be readily performed, in particular in the mouse, and that the effect of therapy on ongoing disease can be more readily assessed.

## Antireceptor Antibodies in Myasthenia and EAMG

### Antibody Levels in Myasthenia

*A) Group Studies.* As pointed out in the initial study of antibody levels in myasthenic patients (Lindstrom et al. 1976c), there is no absolute correlation of titer as assessed in the conventional assay system and signs of myasthenia. Two points stand out:

1. Approximately 90% of patients have circulating antibodies to the nAChR which can be detected using crude preparations of human receptor indirectly labeled with the specific ligand, α-bungarotoxin, i.e., approximately 10% of patients do not have detectable antibodies.
2. The mean level of antibodies is in general higher in patients with more severe disease, but a complete spectrum of titers can be obtained in each group.

#### *Lack of Detectable Circulating Antibody*

Assuming that the diagnosis of myasthenia is reliable, the apparent lack of circulating antibody might be explained in a number of ways:

a) It is possible that the lack of circulating antibody might result from the fixation of the antibody at the endplate. For example, a fall in circulating antibody has been noted in myasthenic patients (Harvey et al. 1978b) and myasthenic chickens (Barkas et al. 1978) during worsening of the symptoms.

b) The type of antigen used might be important, i.e., not all acetylcholine receptors are equal. Differential reactivity of antibodies from patients with receptor isolated from different muscles has been reported (Vincent and Newsom-Davis 1979, Lefvert 1982).
c) The assay system of receptor indirectly labeled with α-toxin does not permit the detection of antibodies to sites covered by the toxin molecule. Some sera which are negative in the conventional assay system were found to block the response of cultured muscle cells to applied acetylcholine (Harvey et al. 1978b). A second drawback to the toxin-labeled receptor system is that the sera of many myasthenic patients appear to contain antibodies which can displace the bound toxin (Barkas and Simpson 1982c).

*Lack of Correlation with Severity*

A lack of correlation might result either from the clinical assessment of severity or from the assay procedure. It must be admitted that it is somewhat difficult to make a quantitative assessment of severity of disease between patients. Secondly, most studies are not carried out on untreated patients, but on those undergoing or having undergone various therapies which might influence the clinical assessment. Some limitations of the assay system have been described above. Others are that:

a) High levels of antibodies in patients with mild disease might be explicable by the specificity or type of antibody. In terms of specificity, some antibodies might be directed against sites exposed on the soluble receptor used for the assay which are not exposed on cell-bound nAChR, others might not be effective in crosslinking receptor molecules which is required for enhancement of the degradation process (see below), others (which would not be detected) might be directed to labile sites lost on receptor preparation. In terms of type of antibody, classes, or subclasses which can preferentially activate complement or direct cytotoxic cells might be relevant.
b) Circulating antibody might be masked in the form of immune complexes. Complexes have been observed in the sera but their specificity has not been adequately tested (Casali et al. 1976, Behan and Behan 1979, Bartolini et al. 1981, Barkas et al. 1984).

*B) Individual Patients.* Fortunately in the case of individual patients, most of the above reservations can be discarded. For example, it is relatively easy to assess improvement of an individual and the specificity of antibody appears relatively stable (Savage-Marengo et al. 1980). Assessment of antibody titers and correlation with severity of disease have been successfully carried out in a number of studies (e.g., Newsom-Davis et al. 1978, Hawkey et al. 1981).

## Effects of Antibodies

It has been clearly shown that anti-nAChR antibody can account for the symptoms of myasthenia. In individual patients, there is a good correlation of titer of antibodies and severity of disease. In EAMG, successful transfer of symptoms has been performed by passive transfer of IgG from myasthenic patients (Toyka et al. 1977), antibodies from myasthenic animals (Lindstrom et al. 1976a) or monoclonal antibodies (e.g., Lennon and Lambert 1980). Neonatal myasthenia involving placental transfer of IgG is a further example. It must be borne in mind, however, that an involvement of cytotoxic cells has never been formally discounted.

Anti-receptor antibodies can be shown in vivo and in vitro to exert their effects in various ways. A rapid block of receptor responses to acetylcholine can be seen using cultured muscle cells (Harvey et al. 1978a,b). A slower effect with a time course of several hours is that of the acceleration of the degradation rate of the receptor produced by cross-linking of receptor molecules, leading to internalization and degradation. This can be demonstrated both in vitro (Anwyl et al. 1977, Bevan et al. 1977) and in vivo (Merlie et al. 1979). Finally, as described above, there is extensive destruction and modification of the endplate region. The exact relevance of each to myasthenia gravis is not clear.

## Structure of the nAChR

### Overall Structure

The structure of the AChR has recently been reviewed (Changeux et al. 1984).

The nAChR is an intrinsic membrane protein which can be solubilized in an active form by nonionic detergents and purified using specific ligands, $\alpha$-toxins from the venoms of snakes, such as the cobra. The receptor purified from *Torpedo* has an apparent molecular weight of 270-290 K daltons on gel filtration. On SDS-gel electrophoresis, four polypeptide bands are seen, $\alpha$, $\beta$, $\gamma$, and $\delta$, present in the ratio of 2:1:1:1 and with apparent molecular weights of 40, 50, 60, and 65 K daltons, respectively. All are glycosylated. The alpha chain bears the binding site for cholinergic ligands and toxins and contains the main immunogenic site of the receptor (see below). The muscle form of other species appears very similar.

The primary sequences of all four chains of *Torpedo* nAChR (Noda et al. 1982, 1983b,c, Sumikawa et al. 1982, Claudio et al. 1983, Devillers-Thiery et al. 1983), the $\alpha$- and $\gamma$-chain of human (Noda et al. 1983a, Shibahara et al. 1985), $\alpha$, $\beta$, and $\gamma$ of calf (Noda et al. 1983a, Takai et al. 1984, Tanabe et al. 1984), $\alpha$ and $\delta$ of mouse (La Polla et al. 1984, Boulter et al. in press) and $\gamma$ and $\delta$ of chick (Nef et al. 1984) nAChRs have been deduced from cDNA and genomic sequences. The molecular weights calculated from the sequences are 50, 54, 57, and 58 K daltons for the protein moiety, the discrepancy between these values and the results from SDS gels not being due to posttranslational cleavage (Barkas et al. 1984), but appearing to result from abnormal behavior of the chains on the gels due to the C-terminal part of the polypeptide (Barkas and Schwendimann, unpublished results). All four chains show extensive homology, pointing to a common evolutionary origin. Across species, the alpha chain is highly conserved with more than 80% identity of residues between human and *Torpedo* receptors. Other chains are less well conserved (50%-60% identity).

Hydrophobicity profiles based on the sequence data also suggest a common means of insertion of the individual chains in the membrane. Four hydrophobic regions, consisting of approximately 20 amino acids and bounded by charged groups, can be seen in each sequence (Fig. 1). Two models have been proposed (Fig. 2). In both, the N-terminal half of

```
              5         10        15        20        25        30
              :         :         :         :         :         :
  1 S E H E T R L V A N L L E N Y N K V I R P V E H H T H F V D
 31 I T V G L Q L I Q L I S V D E V N Q I V E T N V R L R Q Q W
 61 I D V R L R W N P A D Y G G I K K I R L P S D D V W L P D L
 91 V L Y N N A D G D F A I V H M T K L L L D Y T G K I M W T P
121 P A I F K S Y C E I I V T H F P F D Q Q N C T M K L G I W T
151 Y D G T K V S I S P E S D R P D L S T F M E S G E W V M K D
181 Y R G W K H W V Y Y T C C P D T P Y L D I T Y H F I M Q R I
211 P L Y F V V N V I I P C L L F S F L T G L V F Y L P T D S G
241 E K M T L S I S V L L S L T V F L L V I V E L I P S T S S A
271 V P L I G K Y M L F T M I F V I S S I I I T V V V I N T H H
301 R S P S T H T M P Q W V R K I F I D T I P N V M F F S T M K
331 R A S K E K Q E N K I F A D D I D I S D I S G K Q V T G E V
361 I F Q T P L I K N P D V K S A I E G V K Y I A E H M K S D E
391 E S S N A A E E W K Y V A M V I D H I L L C V F M L I C I I
421 G T V S V F A G R L I E L S Q E G
```

PEPTIDES P1, P9 AND P3 ARE RESIDUES 151-169,378-392 AND 426-437.

THE FOUR POSTULATED TRANSMEMBRANE SECTIONS I,II,III AND IV ARE UNDERLINED.

Fig. 1. Sequence of the alpha chain of the nAChR of *Torpedo californica*. (Noda et al. 1982)

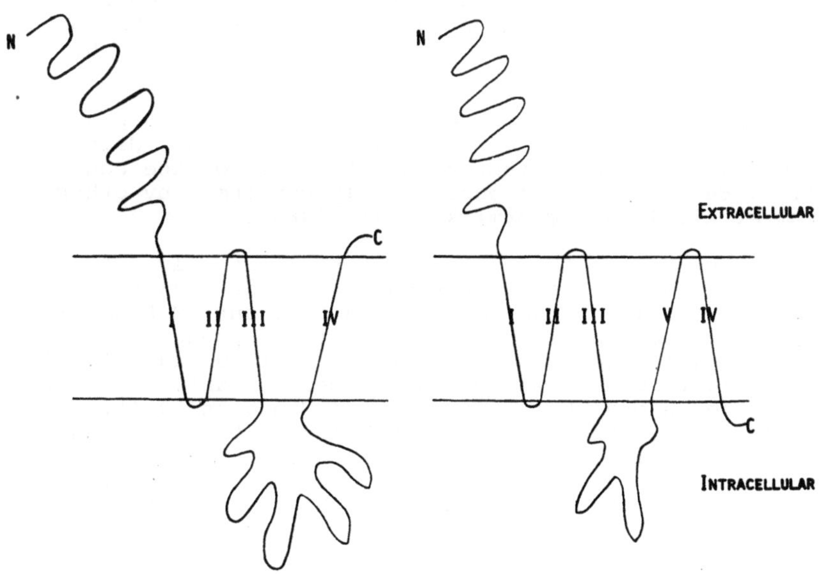


Fig. 2. Models for the insertion of the chains in the membrane

the chain is located extracellularly. This part includes the ligand-binding site, the carbohydrate attachment site, and the main immunogenic region (MIR) in the case of the alpha chain. In one model, each chain crosses the membrane three times, forms a large cytoplasmic loop and finally crosses the membrane once more, placing the C-terminus outside the cell (Claudio et al. 1983, Devillers-Thiery et al. 1983, Noda et al. 1983b). In the second model, a fifth transmembrane region, an amphipathic helix, is inserted between transmembrane regions III and IV, placing the C-terminus inside the cell (Guy, 1984, Finer-Moore and Stroud, 1984). Recent experiments tend to support the latter model (Lindstrom et al. 1984, Young et al. 1985). The proposed amphipathic helix (including peptide P9 in Fig. 2) is suggested as a candidate for the ion-channel of the receptor.

## The Main Immunogenic Region

The term "main immunogenic region" (MIR) was introduced by Tzartos and Lindstrom to describe their observation that most monoclonal antibodies to the nAChRs from *Torpedo* and the electric eel, *Electrophorus*, cross-competed with each other for binding to the receptor (Tzartos and Lindstrom 1980, Tzartos et al. 1981). The majority of antibodies in polyclonal antisera were found to be directed to the same region. This suggests that a relatively small region of the receptor is immunodominant. Many of the anti-MIR antibodies bind to a certain extent to nAChR of other species, enabling the same workers to demonstrate that the majority of antibodies in myasthenic sera are directed against a similar MIR (Tzartos et al. 1982). When isolated chains of nAChR are used as antigen, the alpha chain is preferentially precipitated by antibodies to the MIR.

Based on the hydrophobicity profile, peptide 161-166 of the alpha chain was suggested as a candidate for the MIR (Noda et al. 1982). Using anti-peptide antibodies, we have demonstrated that this is not the case (Juillerat et al. 1984), this being subsequently confirmed by other groups (Lindstrom et al. 1984, McCormick et al. 1985).

Current thinking on the MIR is that it is a conformationally determined site, possibly consisting of a relatively short peptide (Tzartos et al. 1981). This is based firstly on the observation that anti-MIR antibodies bind only weakly to subunits of nAChR in the presence of SDS. More recently, however, Tzartos has demonstrated that, using suitable lipids, it is possible to bind both toxins and certain anti-MIR antibodies to SDS-denatured alpha chain (Tzartos and Changeux 1984). Secondly, it has been reported that, during synthesis of the alpha chain, reactivity with either toxins or anti-MIR antibodies is only observed subsequent to the formation of a disulfide bridge and the addition of carbohydrate, suggesting that a conformational change brought about by these modifications stabilizes the MIR (Merlie et al. 1982). However, in experiments described below, we have preliminary evidence that anti-MIR antibodies bind to a peptide lacking the carbohydrate residues and probably the disulfide bridge which is thought to be formed between residue 128 or 142 and 192 or 193 (Kao et al. 1984, see Fig. 1).

## Approaches to Immunotherapy of EAMG

The usual treatments for myasthenia gravis are the use of anticholinesterase agents and nonspecific immunotherapy, such as thymectomy, cytotoxic drugs, and plasmapheresis. Apart from lack of specificity, such therapeutic regimes can have severe side effects and can be costly. A specific treatment with no side effects would thus be of interest.

In recent years, many immunotherapeutic approaches have been tested using the experimental models of myasthenia (Table 1). While interesting in themselves and providing information on the immune system in general, the majority are obviously only applicable to the experimental model and, in most cases, the specificity of the treatments has not been tested.

Three of the more interesting approaches are the recent production of suppressor cells and factors, the use of toxic antigens, and the possible use of synthetic peptides containing the MIR. These are discussed in detail below.

Table 1. Treatments for EAMG

| Treatment | Effective: | | | Specificity | Reference |
|---|---|---|---|---|---|
| | In vitro | Before symptoms | After symptoms | | |
| Reduced carboxymethylated receptor (immunization) | ? | ± | ± | ? | A |
| DMSO (i.p.) | ? | ? | Ab levels drop | + | B |
| α-fetoprotein (i.p.) | ? | + | + | ? | C |
| Antiidiotype Ab | ? | - | ? | ? | D, E |
| nAChR-Ab complexes (immunization) | ? | + | ? | ? | F |
| Prostaglandin E1 (s.c.) | ? | + | ? | ? | G |
| Electrolectin (i.d. or immunization) | ? | + | + | + | H |
| Anti-IA Ab (i.p.) | ? | + | ? | ? | I |
| Suppressor cells/factors | + | + | ? | ± | J, K, L |
| Toxic antigens | | | | | |
| A Ricin-conjugated | + | ? | ? | ± | M |
| B Radioactive (i.v.) | ? | + | ? | ? | N |
| Synthetic peptides | ? | ? | ? | ? | ? |

*A* Bartfeld and Fuchs, 1978; *B* Pestronk and Drachman, 1980; *C* Brenner and Abramsky, 1981; *D* Lennon and Lambert, 1981; *E* Barkas and Simpson, 1982a; *F* Barkas and Simpson, 1982c; *G* Takamori and Ide, 1982; *H* Levi et al., 1983; *I* Waldor et al., 1983; *J* Bogen et al., 1984; *K* Pachner and Kantor, 1984; *L* Simigaglia et al., 1984; *M* Killen and Lindstrom, 1984; *N* Sterz et al., 1985.

## Suppressor Cells and Factors

Three groups have used this approach (see Table 1). Bogen et al. (1984) induced suppressor cells in mice by intravenous injection of soluble *Torpedo* nAChR. A soluble factor secreted in vitro by these cells was effective in vitro in inhibiting anti-nAChR antibody production and was without effect on an unrelated system. As a second approach, avoiding immunization in vivo, they induced suppressor cells in vitro by culturing spleen cells from naive mice with nAChR in the presence of fetal calf serum (FCS). Again, anti-nAChR responses were inhibited, but there was a certain amount of nonspecificity. Using the supernatants from these cells, similar results were obtained. However, the nonspecific activity could be removed by passage of the supernatants over immobilized FCS. Pachner and Kantor (1984), using a combination of in vivo and in vitro sensitization to *Torpedo* nAChR, obtained nine suppressor cell lines, which were more effective in suppressing in vitro an anti-nAChR response than an anti-ovalbumin response, as assessed by proliferative responses and antibody production. Interestingly, when such cell lines were injected intravenously immediately prior to immunization of the recipients, the production of anti-nAChR antibodies was specifically suppressed for at least 4 weeks and symptoms of myasthenia were markedly reduced. Sinigaglia et al. (1984) immunized mice with *Torpedo* nAChR, removed the popliteal lymph nodes and enriched the percentage of nAChR-specific T cells by in vitro culture with nAChR, followed by removal of B cells. The enriched T cells were infected with the transforming radiation leukemia virus, RadLV, and injected into syngeneic mice. The lymphomas formed were then used to establish cell lines and clones. Supernatants from one line and, surprisingly, from the clones derived from it, inhibited the in vitro proliferative response to *Torpedo* nAChR by 60%-90%, whereas they had no effect on the response to calf nAChR. Culture supernatants from the cell line, when given intravenously following immunization, also suppressed the antibody response to *Torpedo* but not to calf nAChR or α-bungarotoxin for periods of at least 30 days.

## Toxic Antigens

Radiolabeled nAChR given intravenously prior to immunization has been reported to decrease considerably the anti-nAChR response and loss of receptor in rats (Sterz et al. 1985). This study was rather unusual in that 5 doses of 100 μg of nAChR were used per rat. Using nAChR coupled to the toxic protein, ricin, Killen and Lindstrom (1984) were able to find conditions in vitro in which the B cell response to the nAChR could be effectively and specifically suppressed, whereas the T cell proliferative response to the nAChR was not markedly affected. Conditions which did result in depression of the T cell response also resulted in considerable nonspecific cytotoxicity.

## Identification of the MIR and Use of Synthetic Peptides

If the MIR could be localized to one or a few linear sequences of the alpha chain, we could envisage the specific therapeutic use of small peptides either as immunoadsorbents to remove antibodies or antibody-forming cells or, after coupling to an appropriate toxic agent, as cytotoxins directed against antibody-forming cells. Given the primary sequences determined for the receptor, several approaches to the localization of the MIR are possible (Table 2). Experiments using site-directed mutagenesis have been reported, so far without observed effects on antigenicity (Mishina et al. 1985). The approach which we are currently using is that of enzymatic and chemical fragmentation of the

Table 2. Possible approaches to the localization of the MIR

| |
|---|
| 1. Site-directed mutagenesis |
| 2. Shotgun cloning of randomly cut coding sequences into an expression vector |
| 3. Enzymatic and chemical cleavage of the polypeptide chain |

alpha chain, separation of the products by SDS gel electrophoresis and their characterization using antibodies to the MIR and to synthetic peptides, some of which are shown in Fig. 1. Using this approach, it has been possible to localize the MIR to a region N-terminal to peptide P1 (Fig. 3, Barkas, Gabriel, Juillerat, and Tzartos, unpublished results). Digestion of nAChR with papain yields a peptide of 27 K daltons reactive with anti-P1 and anti-P3 antibodies, but not with anti-MIR antibodies or Concanavalin A-Sepharose, and a peptide of 19 K daltons which reacts with anti-MIR antibodies, bears carbohydrate and does not react with anti-P1 or anti-P3. As only glycoproteins linked via asparagine residues are bound to Con A-Sepharose (Krusius 1976) and the only such site in the alpha chain is Asn 141, the papain cleavage site must be between Asn 141 and P1 (151-169), with the MIR N-terminal to P1. This is confirmed using nAChR purified from trypsinized membranes in which the C-terminal 10 K dalton unit is removed (Barkas et al. 1984). In this case, the P1 containing peptide is smaller, whereas the MIR peptide remains the same size. Finally, the P1 containing peptide apparently also contains the C-terminal peptide P3. It is interesting to note that a minor component of less than 15 K daltons molecular weight also reacts with anti-MIR antibodies but does not bind to Con A-Sepharose, suggesting that the MIR is N-terminal to Asn 141 and that the antigenic site is independent of carbohydrate and the original disulfide bridge.

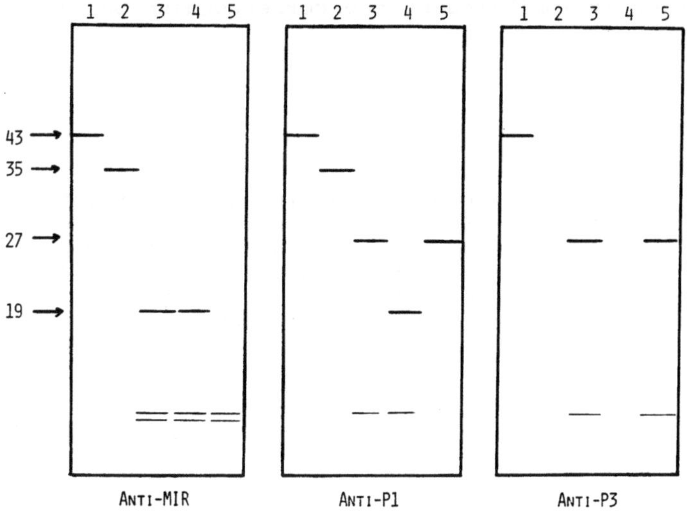


Fig. 3. Schematic representation of immunoblots showing peptides reactive with antibodies to the MIR and synthetic peptides. Lanes *1* nAChR; *2* nAChR purified from trypsinized membranes; *3* and *4* as *1* and *2* but digested with papain; *5* as *3* but after absorption with Con A-Sepharose (12.5% gel)

## Involvement of Other Forms of the Nicotinic Acetylcholine Receptor

It has been suggested that anti-nAChR antibodies in binding to the CNS form of the nAChR might be responsible for CNS disturbances in myasthenic patients (Fulpius et al. 1977, Lefvert and Pirskanen 1977, Fontana et al. 1978). As described above, the muscle form of the nAChR has been exhaustively characterized. In contrast, little is definitely known about the CNS nAChR and most of this is contradictory. Binding of the "specific" ligand, α-bungarotoxin, to CNS tissue is observed, but most evidence suggests that this is not to the functional nAChR (reviewed by Merlie et al. 1979). At least in the case of the rat sympathetic cell line, PC 12, it is clear that the toxin-binding component, which is antigenically unrelated to the muscle nAChR, is distinct from the functional nAChR, which is antigenically related to the muscle nAChR (Patrick and Stallcup 1977). Partial (Betz 1981, Wonnacott et al. 1982) or complete (Norman et al. 1982) purification of a toxin-binding component from rat and chick CNS have been reported; these proteins are reported as antigenically related to the muscle form. In contrast, Lindstrom et al. (1983) have partially purified a nontoxin binding component from chick brain; this also is antigenically related to the muscle nAChR. Using affinity-purified anti-peptide antibodies to the "amphipathic helix" (peptide P9, Fig. 1), we have observed a strong reaction with a component of chick brain membranes with an apparent molecular weight of 52 K daltons (Fig. 4, Barkas and Juillerat, unpublished results). Recently, a chick genomic clone has been isolated and sequenced which shows 50% identity of residues at the amino acid level with the alpha chain of chick muscle nAChR and is more related to the alpha chain than to the other chains. The calculated molecular weight would be approximately 50 K daltons (Ballivet, unpublished results). Antipeptide antibodies to the equivalent of the P1 peptide from this sequence (a region which shows no homology whatsoever to the muscle alpha form) also recognize a protein in chick brain membranes of apparent molecular weight 49 K daltons (Fig. 4, our unpublished results). In both cases, the binding of the affinity-purified antibodies is specifically inhibited by the corresponding peptide (Fig. 4).

## Conclusion

Remarkable progress has been made in the understanding and treatment of myasthenia gravis in the last 50 years. Major milestones have been the realization that the weakness was due to deficiencies at the neuromuscular junction and that the disease might be autoimmune, the demonstration of the involvement of the nAChR and, more recently, the structural work. Potential approaches to immunotherapy are those of classical immunology and the use of synthetic antigens. Finally, the story is clearly not finished, as it is becoming increasingly clear that a family of nAChR-like molecules exist.

*Acknowledgments.* Financial support for our work was provided by the Swiss National Fund for Research (grants nos. 3.450.0.83 and 3.432.0.83).

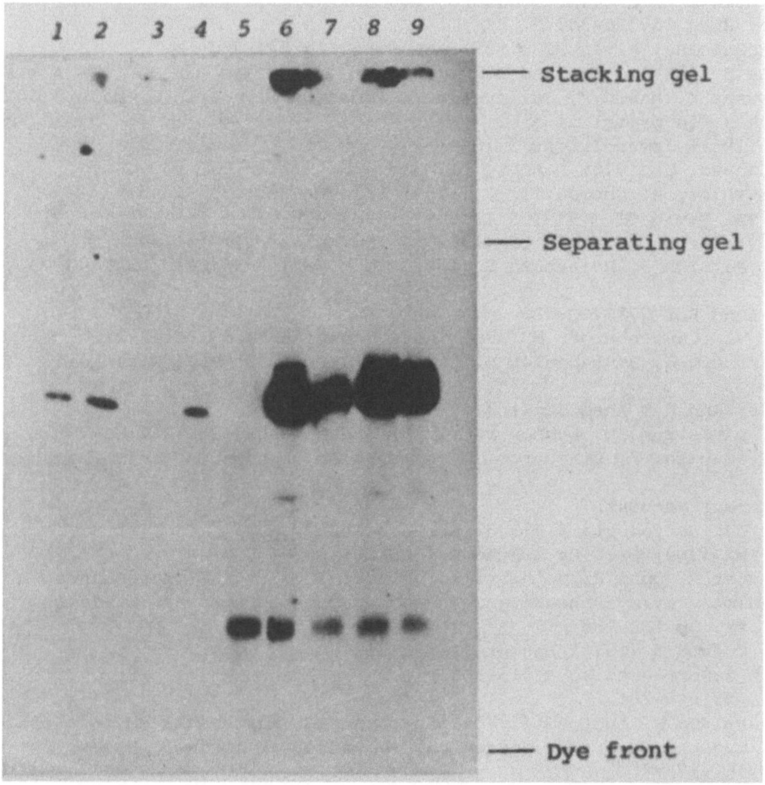


Fig. 4. Immunoblot of embryonic chick brain membranes. Affinity-purified antibodies to peptides CP2 (lanes *1-4*) or P9 (lanes *6-9*) or normal IgG (lane *5*) at equivalent concentrations were incubated alone (lanes *1*, *5*, *6*) or with peptides P9 (lanes *2*, *7*), CP2 (lanes *3*, *8*) or CP1 (lanes *4*, *9*) before incubation with the separated brain proteins transferred to nitrocellulose. CP1 is the peptide in chick alpha chain and CP2 that in the chick "alpha-like" sequence equivalent to P1 in the alpha chain of *Torpedo* (10% gels)

## References


Anwyl R, Appel SM, Narahashi T (1977) Nature (London) 267:262-263

Barkas T, Boyle RS, Behan PO (1981) J Clin Lab Immunol 5:27-30

Barkas T, Simpson JA (1982a) Clin Exp Immunol 47:119-126

Barkas T, Simpson JA (1982b) J Clin Lab Immunol 7:223-227

Barkas T, Simpson JA (1982c) J Clin Lab Immunol 9:113-117

Barkas T, Harrison R, Lunt GG, Watson CMJ, Harvey AL, Robertson JG (1978) Biochem. Soc Trans 6:634-636

Barkas T, Juillerat M, Kistler J, Schwendimann B, Moody J (1984) Eur J Biochem 143: 309-314

Bartfeld D, Fuchs S (1978) Proc Natl Acad Sci USA 75:4006-4010

Bartolini C, Guidi L, Scoppetta C, Tonali P, Bartoccioni E, Flamini G, Gambassi G, Terranova T (1981) J Neurol Neurosurg Psychiatry 44:901-905

Behan WMH, Behan PO (1979) J Neurol Neurosurg Psychiatry 42:595-599

Berman P, Patrick J (1980a) J Exp Med 151:204-223

Berman P, Patrick J (1980b) J Exp Med 152:507-520

Betz H (1981) Eur J Biochem 117:131-139

Bevan S, Kullberg RW, Heinemann SF (1977) Nature (London) 267:263-265
Blalock A (1944) J Thor Surg 13:316-339
Blalock A, Mason MF, Morgan HJ, Riven SS (1939) Ann Surg 110:544-560
Bogen S, Mozes E, Fuchs S (1984) J Exp Med 159:292-304
Boulter J, Luyten W, Evans K, Mason P, Ballivet M, Goldman D, Stengelin S, Martin G, Heinemann S, Patrick J (in press)
Brenner T, Abramsky O (1981) Immunol Lett 3:163-167
Casali P, Borzini P, Zanussi C (1976) Lancet, II, 375
Changeux JP, Devillers-Thiery A, Chemouilli P (1984) Science 225:1335-1345
Christadoss P, Lennon VA, David CS (1979) J Immunol 123:2540-2543
Christadoss P, Lennon VA, Krco CJ, David CS (1982) J Immunol 128:1141-1144
Claudio T, Ballivet M, Patrick J, Heinemann S (1983) Proc Natl Acad Sci (USA) 80: 1111-1115
Dau PC (1981) Ann N Y Acad Sci 377:700-708
Delwaide PJ, Salmon J, Van Cauwenberger H (1967) Acta Neurol Belg 67:701-712
Devillers-Thiery A, Giraudat J, Bentaboulet M, Changeux J-P (1983) Proc Natl Acad Sci USA 80:2067-2071
Engel AG, Santa T (1971) Ann N Y Acad Sci 183:46-63
Engel AG, Tsujihata M, Lindstrom JM, Lennon VA (1976) Ann N Y Acad Sci 274:60-79
Engel AG, Sakakibara H, Sahashi K, Lindstrom JM, Lambert EH, Lennon VA (1979) Neurology 29:179-188
Erb W (1879) Arch Psychiatr Nervenkr
Fambrough DM, Drachman DB, Satyamurti S (1973) Science 182:293-295
Finer-Moore J, Stroud RM (1984) Proc Natl Acad Sci USA 81:155-159
Fontana A, Grob PJ, Sauter R, Dubs R, Mothersill I (1978) In: Lunt GG, Marchbanks RM (eds) The biochemistry of myasthenia gravis and muscular dystrophy. Academic Press, London New York, pp 183-188
Fulpius BW, Fontana A, Cuenod S (1977) Lancet II:350-351
Goldflam S (1893) Dtsch Z Nervenheilkd 4:313-352
Guy HR (1984) Biophys J 45:249-261
Harvey AL, Barkas T, Harrison R, Lunt GG (1978a) In: Lunt GG, Marchbanks RM (eds) The biochemistry of myasthenia gravis and muscular dystrophy. Academic Press, London New York, pp 167-175
Harvey AL, Robertson JG, Barkas T, Harrison R, Lunt GG, Stephenson FA, Campbell MJ, Teague RH (1978b) Clin Exp Immunol 34:411-416
Hawkey CJ, Newsom-Davis J, Vincent A (1981) J Neurol Neurosurg Psychiatry 44:469-475
Jolly F (1895) Berl Klin Wochenschr 32:1-7
Juillerat MA, Barkas T, Tzartos SJ (1984) FEBS Lett 168:143-148
Kao PN, Dwork AJ, Kaldany R-RJ, Silver ML, Wideman J, Stein S, Karlin A (1934) J Biol Chem 259:11662-11665
Killen JA, Lindstrom JM (1984) J Immunol 133:2549-2553
Krusius T (1976) FEBS Lett 66:86-89
La Polla RJ, Mayne KM, Davidson N (1984) Proc Natl Acad Sci USA 81:7970-7974
Lefvert AK (1982) J Neurol Neurosurg Psychiatry 45:70-73
Lefvert AK, Pirskanen R (1977) Lancet II:352
Lennon VA, Lambert EH (1980) Nature (London) 285:238-240
Lennon VA, Lambert EH (1981) Ann N Y Acad Sci 377:77-95
Lennon VA, Lindstrom JM, Seybold ME (1975) J Exp Med 141:1365-1375
Lennon VA, Lindstrom JM, Seybold ME (1976) Ann N Y Acad Sci 274:283-299
Lennon VA, Seybold ME, Lindstrom JA, Cochrane C, Ulevitch R (1978) J Exp Med 147: 973-982
Levi G, Tarrab-Hazdai R, Teichberg VI (1983) Eur J Immunol 13:500-507
Lindstrom JM, Engel AG, Seybold ME, Lennon VA, Lambert EH (1976a) J Exp Med 144: 739-753
Lindstrom JM, Lennon VA, Seybold ME, Whittingham S (1976b) Ann N Y Acad Sci 274: 254-274
Lindstrom JM, Seybold ME, Lennon VA, Whittingham S, Duane DD (1976c) Neurology 26: 1054-1059
Lindstrom JM, Tzartos S, Gullick W, Hochschwender S, Swanson L, Sargent P, Jacob M, Montal M (1983) Cold Spring Harbor Symp Quant Biol XLVIII:89-99
Lindstrom JM, Criado M, Hochschwender S, Fox JL, Sarin V (1984) Nature (London) 311:573-575

McCormick DJ, Lennon VA, Atassi MZ (1985) Biochem J 226:193-197
Merlie JP, Heinemann S, Einarson B, Lindstrom JM (1979) J Biol Chem 254:6328-6332
Merlie JP, Sebbane R, Tzartos S, Lindstrom JM (1982) J Biol Chem 257:2694-2701
Mishina M, Tobimatsu T, Imoto K, Tanaka K-I, Fujita Y, Fukuda K, Kurasaki M, Takahashi H, Morimoto Y, Hirose T, Inayama S, Takahashi T, Kuno M, Numa S (1985) Nature (London) 313:364-369
Morley BJ, Kemp GE, Salvaterra P (1979) Life Sci 24:859-872
Nastuk WL, Strauss AJL, Osserman KE (1959) Am J Med 26:394-409
Nastuk WL, Plescia OJ, Osserman KE (1960) Proc Soc Exp Biol Med 105:177-184
Nef P, Mauron A, Stalder R, Alliod C, Ballivet M (1984) Proc Natl Acad Sci USA 81: 7975-7979
Newsom-Davis J, Pinching AJ, Vincent A, Wilson SG (1978) Neurology 28:266-272
Noda M, Takahashi H, Tanabe T, Toyosato M, Furutani Y, Hirose T, Asai M, Inayama S, Miyata T, Numa S (1982) Nature (London) 299:793-797
Noda M, Furutani Y, Takahashi H, Toyosato M, Tanabe T, Shimizu S, Kikyotani S, Kayano T, Hirose T, Inayama S, Numa S (1983a) Nature (London) 305:818-823
Noda M, Takahashi H, Tanabe T, Toyosato M, Kikyotani S, Furutani Y, Hirose T, Takashima H, Inayama S, Miyata T, Numa S (1983b) Nature (London) 302:528-532
Noda M, Takahashi H, Tanabe T, Toyosato M, Kikyotani S, Hirose T, Asai M, Takashima H, Inayama S, Miyata T, Numa S (1983c) Nature (London) 301:251-255
Norman RI, Mehraban F, Barnard EA, Dolly JO (1982) Proc Natl Acad Sci USA 79:1321-1325
Oosterhuis HJGH (1984) In: Clinical neurology and neurosurgery monographs, vol V. Churchill Livingstone, Edinburgh London Melbourne New York
Pachner AR, Kantor FS (1984) Clin Exp Immunol 56:659-668
Patrick J, Lindstrom JM (1973) Science 180:871-872
Patrick J, Stallcup WB (1977) Proc Natl Acad Sci USA 74:4689-4692
Pestronk A, Drachman DB (1980) Nature (London) 288:733-734
Sahashi K, Engel AG, Lamberg EH, Howard FM (1980) J Neuropathol Exp Neurol 29:160-172
Savage-Marengo T, Harrison R, Lunt GG, Behan PO (1980) J Neurol Neurosurg Psychiatry 43:316-320
Schumacher CH, Roth P (1912) Mitt Grenzgeb Med Chir 25:746-765
Shibahara S, Kubo T, Perski HJ, Takahashi H, Noda M, Numa S (1985) Eur J Biochem 146:15-22
Simpson JA (1960) Scot Med J 5:419-436
Sinigaglia F, Gotti C, Castagnoli PR, Clementi F (1984) Proc Natl Acad Sci USA 81: 7569-7573
Sterz RKM, Biro G, Rajki K, Filipp G, Peper K (1985) J Immunol 134:841-846
Sumikawa K, Houghton M, Smith JC, Bell L, Richards BM, Barnard EA (1982) Nucleic Acids Res 10:5809-5822
Takai T, Noda M, Furutani Y, Takahashi H, Notake M, Shimizu S, Kayano T, Tanabe T, Tanaka K-I, Hirose T, Inayama S, Numa S (1984) Eur J Biochem 143:109-115
Takamori M, Ide Y (1982) Neurology 32:410-413
Tanabe T, Noda M, Furutani Y, Takai T, Takahashi H, Tanaka K-I, Hirose T, Inayama S, Numa S (1984) Eur J Biochem 144:11-17
Toyka KV, Drachman DB, Griffin DE, Pestronk A, Winkelstein JA, Fischbeck KH, Kao I (1977) N Engl J Med 296:125-131
Tzartos SJ, Changeux J-P (1984) J Biol Chem 259:11512-11519
Tzartos SJ, Lindstrom JM (1980) Proc Natl Acad Sci USA 77:755-759
Tzartos SJ, Rand DE, Einarson BL, Lindstrom JM (1981) J Biol Chem 256:8635-8645
Tzartos SJ, Seybold ME, Lindstrom JM (1982) Proc Natl Acad Sci USA 79:188-192
Vincent A, Newsom-Davis J (1979) Adv Cytopharmacol 3:269-278
Von Reis G, Liljestrand A, Matell G (1963) Acta Neurol Scand 41:Suppl 13, part II, 463-471
Waldor MK, Sriram S, McDevitt HO, Steinman L (1983) Proc Natl Acad Sci USA 80:2713-2717
Walker MB (1934) Lancet I:1200-1201
Walker MB (1935) Proc R Soc Med 28:33-35
Warmolts JR, Engel WK, Whitaker JN (1970) Lancet II:1198-1199
Weigert C (1901) Neurol Zentralbl 20-597

Willis T (1672) De anima brutorum. Oxford, pp 404-406
Wonnacott S, Harrison R, Lunt GG (1982) J Neuroimmunol 3:1-13
Young EF, Ralston E, Blake J, Ramachandran J, Hall ZW, Stroud RM (1985) Proc Natl Acad Sci USA 82:626-630

# Biochemistry of Gangliosidoses

K. Sandhoff and E. Conzelmann[1]

## Sphingolipid Metabolism and Storage Diseases

Lipid storage disorders comprise a rather heterogeneous group of progredient and often fatal diseases which mainly disable the nervous system. The biochemical analysis of these diseases led to the discovery of several glycosphingolipids and triggered the investigation of their metabolism.

Glycosphingolipids are characteristic components of the outer leaflets of plasma membranes of animal cells which form cell type specific glycolipid patterns that change with cell differentiation and cell transformation (Hakomori 1975). The physiological function of glycosphingolipids is still unclear, but gangliosides, the sialic acid-containing glycosphingolipids, have been recognized as binding sites for toxins and viruses (Yamakawa and Nagai 1978, Markwell et al. 1981) and as modulators of neuritogenesis and axonal sprouting (Di Gregorio et al. 1984).

Biosynthesis of sphingolipids is catalyzed by a group of membrane-bound transferases mainly localized in the endoplasmic reticulum and the Golgi apparatus (Schachter and Roseman 1980). Within the plasma membrane some sphingolipids such as sphingomyelin and oligosialogangliosides can be modified by plasma membrane-based enzymes which are regulated by the organization and by the physical properties of the membranes (Sandhoff 1984a).

Catabolism of sphingolipids was studied intensively in many laboratories in the last 20 years (Stanbury et al. 1983, Sandhoff 1977, Sandhoff and Christomanou 1979) in order to understand the molecular basis of the inherited lipidoses and their steadily increasing heterogeneity.

For their final degradation, the sphingolipids are transported into the lysosomal compartment, in a still poorly understood way. Here they are degraded in an acidic environment by exo-hydrolases in a stepwise manner starting at the hydrophilic end of the molecules. The sequence of catabolic steps was elucidated by in vitro studies (Fig. 1). As indicated in Figure 1, almost each of the degrading hydrolases can be deficient in a human lipid storage disease. The principle of these diseases is simple: the inherited deficiency of one of these ubiquitously occurring hydrolases causes the lysosomal storage of its substrates. Nevertheless the diseases resulting from these defects are rather heterogeneous from the biochemical as well as from the clinical point of view (Stanbury et al. 1983).

---

[1]Institut für Organische Chemie und Biochemie der Universität Bonn, Gerhard-Domagk-Straße 1, D-5300 Bonn 1, FRG

36. Colloquium - Mosbach 1985
Neurobiochemistry

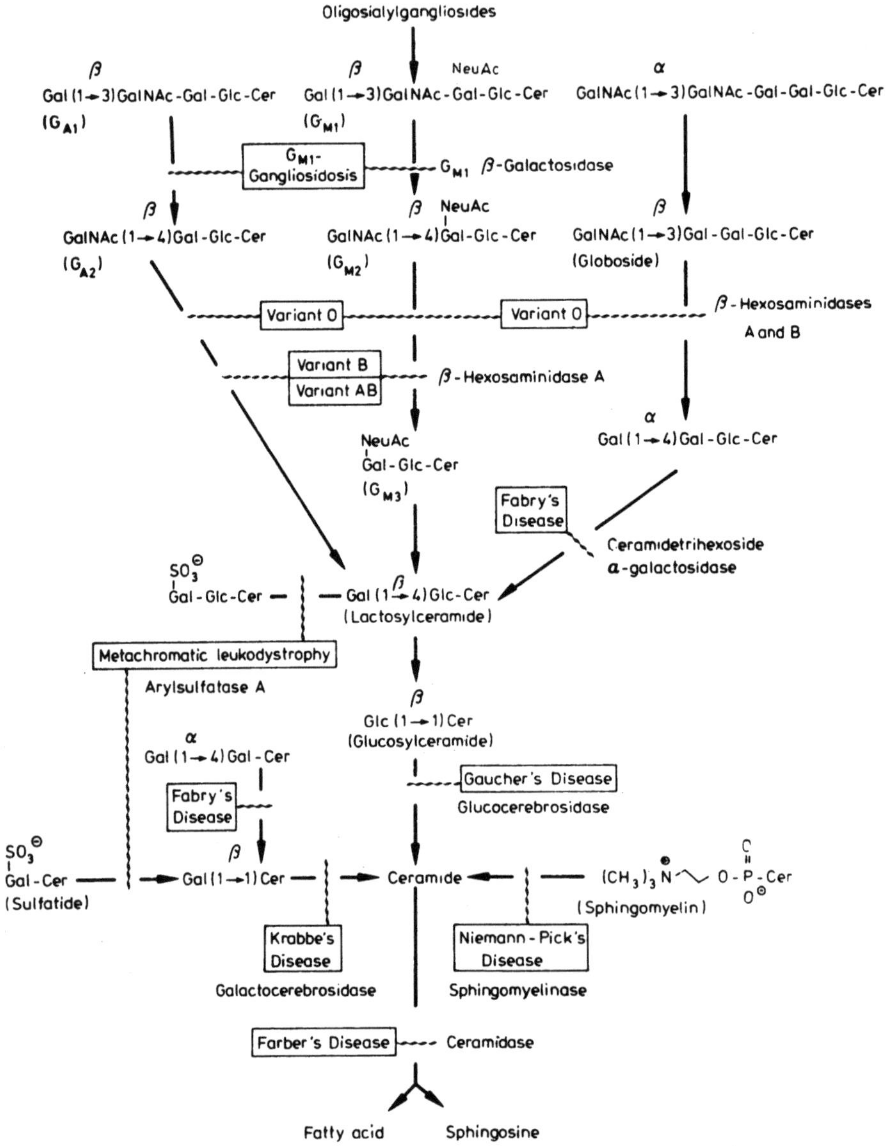


Fig. 1. Degradation scheme of sphingolipids denoting metabolic blocks of known diseases. (Sandhoff and Christomanou 1979). *Cer* ceramide; *Gal* D-galactose, *GalNAc* 2-acetamido-2-deoxy-D-galactopyranoside, *Glc* D-glucose, *NeuAc* N-acetylneuraminic acid. Variant B of infantile $G_{M2}$-gangliosidosis, Tay-Sachs disease; Variant O of infantile $G_{M2}$-gangliosidosis, Sandhoff disease, Sandhoff-Jatzkewitz disease; Variant O of juvenile $G_{M2}$-gangliosidosis, juvenile Sandhoff disease; Variant AB, Variant AB of infantile $G_{M2}$-gangliosidosis

Though the lysosomal hydrolases as well as their inherited defects occur in all organs, body fluids and cells (with the exception of the erythrocytes) lipid accumulation is predominant in those organs where the respective lipid substrates are mainly synthesized. Therefore defects in the ganglioside metabolism will mainly lead to a neuronal storage of glycolipids, whereas defects in sulfatide or galactosylceramide degradation will mainly affect the white matter of the brain and result in leukodystrophies.

## Processing of Lysosomal Enzymes and Causes of Lysosomal Protein Deficiencies

The sphingolipidoses belong to the group of lysosomal storage diseases (Hers 1965), which include defects of glycosidases, sulfatases, and lipid degrading hydrolases. So far no deficiencies of phospholipases, nucleases, or proteinases have been reported. Recently studies on the biosynthesis, post-translational modification and intracellular transport of lysosomal enzymes in normal and mutant cells revealed various causes of lysosomal enzyme deficiencies (von Figura and Hasilik 1984).

Biosynthesis of lysosomal enzymes was studied so far almost exclusively in cultured normal and mutant fibroblasts. Studies with other cell types, especially with brain cells, are still missing. The general picture that emerged indicates that lysosomal enzymes are synthesized like secretory and membrane glycoproteins by ribosomes of the rough endoplasmic reticulum, and are cotranslationally transferred into the lumen of the endopolasmic reticulum (ER) (Erickson and Blobel 1979, Hasilik 1980, Sly and Fischer 1982). After release of an N-terminal signal peptide and attachment of high mannose oligosaccharide cores to some asparagine residues of the nascent polypeptide chain, the lysosomal enzyme precursors are transported from the lumen of the ER into the Golgi complex. After trimming of the carbohydrate chains by specific glycosidases, terminal mannose residues of the lysosomal protein precursors are phosphorylated (for review see: Kornfeld et al. 1983). In fibroblasts the mannose-6-phosphate residues are recognized by a receptor which carries them into the prelysosomal compartment. Here or in the lysosomes the precursors are proteolytically processed to the mature enzymes which persist in the lysosomes with half-lives of several days or more.

A severe reduction of the activity of an enzyme in the lysosome may be caused by any one of a number of inherited defects (for ref. see von Figura and Hasilik 1984). A mutation in the structural gene of the enzyme itself may lead to the synthesis of a mutant enzyme that has a reduced molecular activity or is a poor substrate for post-translational processing or intracellular transport. The defect of an enzyme involved in processing and targeting of lysosomal proteins may cause a multiple enzyme deficiency, as in mucolipidoses II and III, by misdirecting newly synthesized lysosomal enzymes to secretion. In the digestive compartment, the lysosome, the stability of the enzymes against proteolysis is also critical. This stability may be lost, either due to a structural mutation of the enzyme itself, as in some late-onset forms of metachromatic leukodystrophy (von Figura et al. 1983), or owing to the defect of a nonenzymic protein cofactor which normally stabilizes the protein, as in combined β-galactosidase/α-neuraminidase deficiency (D'Azzo et al. 1982). Finally, some water-soluble hydrolases require for their activity towards glycolipid substrates the aid of specific non-enzymic protein cofactors, so-called activator proteins (Sandhoff 1984b). Two lipidoses, i.e., variant AB of $G_{M2}$ gangliosidosis (Conzelmann and Sandhoff 1978) and a juvenile variant of metachromatic leukodystrophy (Stevens et al. 1981), where shown to be caused by a deficiency of the respective activator protein.

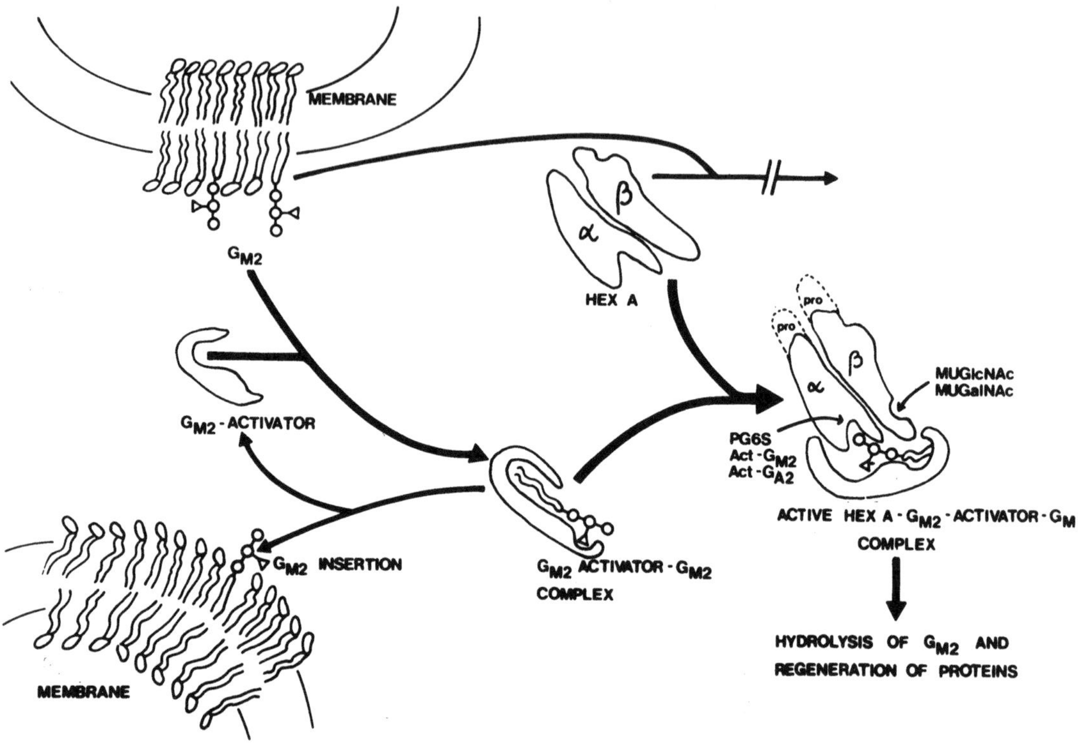


Fig. 2. Model for the lysosomal catabolism of ganglioside $G_{M2}$. Hexosaminidase A cannot attack membrane-bound ganglioside $G_{M2}$. Instead, the ganglioside is extracted from the membrane by the activator protein and the (water-soluble) activator/lipid complex is the substrate for the enzymic reaction. Of the two catalytic sites on hexosaminidase A, only the one on the α-subunit cleaves ganglioside $G_{M2}$. The hexosaminidase precursor ("prohex A") is also fully active on the activator/$G_{M2}$ complex (Hasilik et al. 1982). After the reaction, the product, ganglioside $G_{M3}$, is reinserted into the membrane and the activator protein is available for another round of catalysis. (Conzelmann and Sandhoff 1979, Sandhoff and Conzelmann 1984, Conzelmann et al. 1982)

## Biochemical Basis of $G_{M2}$-Gangliosidoses

All protein defects identified so far in various forms of $G_{M2}$-gangliosidoses directly affect the hydrolysis of ganglioside $G_{M2}$, the main neuronal storage compound. Two proteins are involved in its lysosomal degradation: hexosaminidase A and the $G_{M2}$-activator protein. From the results of in vitro studies we assume that the $G_{M2}$-activator, a lysosomal glycoplipid-binding protein, recognizes membrane-bound ganglioside $G_{M2}$ and solubilizes it as a water-soluble complex (Conzelmann et al. 1982). This complex, which has a molar stoichiometry of 1 : 1, has been isolated by several techniques. Recently, it has been directly demonstrated by affinity labeling of the $G_{M2}$-activator protein with tritium labeled bromoacetyl-lysoganglioside $G_{M2}$ (Neuenhofer and Sandhoff 1985). The activator-lipid complex is recognized as substrate by hexosaminidase A (Hex A) (Fig. 2). After hydrolytic removal of the terminal N-acetylgalactosamine from ganglioside $G_{M2}$, the ganglioside $G_{M3}$-activator complex is released from the enzyme and dissociates,due to the low affinity between product and activator. The free activator can then be reused for further catalytic cycles.

Hex A consists of two nonidentical subunits, α and β[2] (Hasilik and Neufeld 1980), both of which carry an active site (Kytzia et al. 1983). Recent kinetic data indicate that ganglioside $G_{M2}$ and a water-soluble N-acetyl-β-D-glucosaminide-6-sulfate are almost exclusively degraded by the catalytic site on the α subunit whereas the catalytic site on the β-subunit predominantly hydrolyses water-soluble N-acetyl-glucosaminides and -galactosaminides. Ganglioside $G_{M2}$ is recognized by Hex A only after addition of activator protein or, in vitro, of appropriate detergents. On the other hand, the activator inhibits the hydrolysis of sulfated N-acetyl-β-D-glucosaminides by α-sites of Hex A and by the homopolymer $\alpha_2$, Hex S, but not the hydrolysis of N-acetyl-β-D-glucosaminides by the β-site of Hex A and of the homodimer $\beta_2$, Hex B (Kytzia and Sandhoff 1985). These data suggest that the $G_{M2}$-activator specifically recognizes the α-subunit in Hex A and in Hex S. In vitro studies indicate that in the presence of $G_{M2}$ activator, ganglioside $G_{M2}$ is degraded almost exclusively by the catalytic site on the α-subunit of Hex A (Kytzia and Sandhoff 1985), whereas Hex S ($\alpha_2$) and Hex B ($\beta_2$) remain essentially inactive against this lipid substrate. Mutations in all three polypeptides (α-subunit, β-subunit and $G_{M2}$ activator) are known to impair ganglioside $G_{M2}$ catabolism). Thus three biochemical variants of $G_{M2}$-gangliosidosis can be differentiated:

a) Mutations in the α-subunit (encoded on chromosome 15) may lead to the various forms of Tay-Sachs disease (variant B of $G_{M2}$-gangliosidosis). In fibroblasts of most infantile patients analyzed, no material cross-reacting with the α-subunit could be demonstrated with immunochemical techniques (for ref. see Sandhoff and Conzelmann 1984). As a consequence, hexosaminidases A and S are absent. In fibroblasts of one patient, material cross-reactive with monospecific antisera against the α-chain was found in the endoplasmic reticulum or Golgi, but no antigen corresponding to Hex A was detected (Proia and Neufeld 1982). In juvenile and adult patients a defective hexosaminidase A with a residual activity below 5% of normal (Conzelmann et al. 1983) is formed. Recently a new type of B-variant has been recognized. In these patients, a Hex A was found to be active against 4-methylumbelliferyl-β-D-N-acetyl-glucosaminide and -galactosaminide but inactivate against p-nitrophenyl-N-acetyl-β-D-glucosaminide-6-sulfate [PG-6-S, a substrate that is hydrolyzed predominantly by Hex A, practically not by Hex B (Kresse et al. 1981)] and ganglioside $G_{M2}$ (Li et al. 1981, Kytzia et al. 1983).

   Formation and processing of the heteropolymer Hex A in cultured human fibroblasts starts from precursors of the α- and β-chains which are formed in the endoplasmic reticulum (Table 1) (Proia and Neufeld 1982). After attachment of mannose 6-phosphate recognition markers to both polypeptides, they associate probably in the Golgi to yield pro-Hex A (pro α-/pro β-complex) (Proia et al. 1984). This proenzyme is already active against ganglioside $G_{M2}$ and synthetic substrates (Hasilik et al. 1982). It is translocated into the lysosomes and proteolytically clipped to form mature Hex A. In two patients with adult and chronic Tay-Sachs disease, respectively, a defect in the association of pro α- and pro β-chains has been identified (Neufeld et al. 1984). Cultured fibroblasts from these patients produce α- and β-chains and Hex B, but are defective in the production of Hex A and Hex S.

---

[2] Mature β-unit is usually proteolytically processed, see Table 1.

Table 1. Modification of β-hexosaminidase subunits (apparent molecular weight × $10^{-3}$). (Neufeld et al. 1984)

| Form | α-Chain | β-Chain | Where found |
|---|---|---|---|
| Pre-precursor | 65 | 59 | Cell free translation |
| Precursor | 67 | 63 (61) | ER, Golgi secretions |
| Intermediate | 56 | 52 | Lysosomes (?), cell-free proteolysis |
| Mature | 54 | 29 + [24, 22, 19] | Lysosomes |

b) Mutations in the β-subunit (encoded on chromosome 5) lead to various forms of variant 0 of $G_{M2}$-gangliosidosis (Sandhoff's disease). In fibroblasts of most infantile patients analyzed, no β-subunit can be detected by monospecific antisera (Carroll and Robinson 1973, Geiger et al. 1977). Consequently, Hex A as well as Hex B (a homopolymer of β-subunit) are missing (Sandhoff et al. 1968, 1971). In the fibroblasts of one patient, cross-reactive material has been found (Gautron et al. 1983). In juvenile and adult patients, defective Hex A, possessing a small but detectable residual activity, is formed. Mutations of the β-locus have been described that result in the formation of abnormal heat-labile but functionally active hexosaminidases (Navon et al. 1985).

c) In the extremely rare AB variant of $G_{M2}$ gangliosidosis, the $G_{M2}$ activator (encoded on chromosome 5) (Burg et al. 1985) is deficient (Conzelmann and Sandhoff 1978) whereas hexosaminidases A and B appear to be normal (Conzelmann et al. 1978). In the fibroblasts of the few patients analyzed so far, no cross-reactive material corresponding to the mature $G_{M2}$-activator was found (Banerjee et al. 1984).

The $G_{M2}$-activator is a lysosomal glycoprotein with a molecular weight of about 21,500. Besides this intracellular mature form, a higher molecular weight form ($M_r \sim 24{,}000$), which is excreted from cultured human fibroblasts, was detected with polyclonal antibodies. Both forms are deficient in the fibroblasts from patients with variant AB of $G_{M2}$-gangliosidosis analyzed so far. These cells export, however, into the medium small amounts of a 26,000 molecular weight form which can also stimulate the degradation of ganglioside $G_{M2}$ by Hex A in vitro, albeit less efficiently than the mature form of the normal activator.

The physiological importance of the $G_{M2}$-activator for the catabolism of ganglioside $G_{M2}$ was also demonstrated in cell culture (Fig. 3). When ganglioside $G_{M2}$ is added to the culture medium, variant AB cells, like the other variants and normal controls, take up the lipid but, like the enzyme-deficient fibroblasts from patients with variants B and 0, they can only use it to a small extent as precursor for the synthesis of higher gangliosides but are unable to degrade it. Feeding of $G_{M2}$-activator to AB-variant fibroblasts restores their capability to catabolize exogenously added ganglioside $G_{M2}$ to almost normal (Sonderfeld et al. 1985).

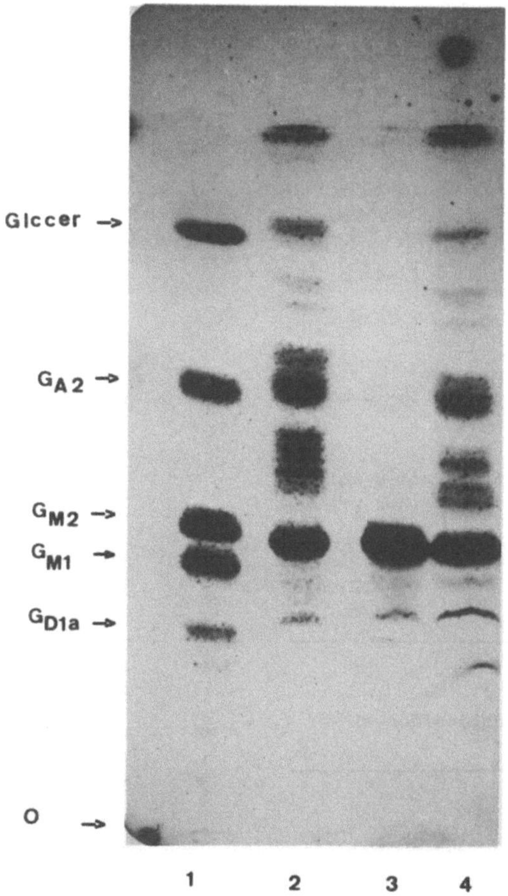


Fig. 3. Metabolism of $[^3H]G_{M2}$ in $G_{M2}$ gangliosidosis, variant AB, cells as function of exogenously added $G_{M2}$-activator protein. The fibroblasts were fed with $[^3H]G_{M2}$ ($5 \times 10^{-5}$ M) and the indicated amounts of purified $G_{M2}$-activator protein. After 70 h, the cells were harvested and processed as described. (Sonderfeld et al. 1985). TLC separation of total lipid extracts: *Lane 1* standards $G_{D1a}$, $G_{M1}$, $G_{M2}$, $G_{A2}$, GlcCer; *Lane 2* total lipid extracts of normal cells; *Lane 3* total lipid extracts of $G_{M2}$ gangliosidosis, variant AB, cells; *Lane 4* total lipid extracts of $G_{M2}$ gangliosidosis, variant AB, cells, after feeding 30 µg $G_{M2}$-activator protein for 70 h; *O* origin

## Phenotypic Description of Tay-Sachs Disease (TSD) and Its Allelic Late Onset Forms

TSD is the prototype of the fatal gangliosidoses. Its heterozygote frequency is 1 in 27 among Ashkenazy Jewish persons but much lower in other populations. Patients develop motor weakness and startle response between 3 to 6 months of age. Mental and motor deterioration progress rapidly after the first year of life. Deafness, blindness and convulsions develop, spasticity appears and decerebrate rigidity is reached. The patients usually die by three years of age. Cherry-red spots in the macular region develop in over 95% of the patients. The cytoplasm of most neurons is ballooned and distended; here multilaminated storage bodies accumulate. They are rich in acid phosphatase and comprise large quantities of the main storage compound, ganglioside $G_{M2}$. Central demyelination appears to be secondary to axonal degeneration and nerve cell death. Cortical gliosis is prominent (O'Brien 1983).

The development of CNS dysfunctions for infantile and juvenile forms are summarized in Tables 2 and 3, respectively. Table 4 lists clinical pictures of adult types.

Table 2. Development of CNS dysfunction in early onset $G_{M2}$ gangliosidoses. (Schulte 1984)

| | | |
|---|---|---|
| Excitability: | | |
| Listlessness-irritability | | |
| Hyperacuses with ext. response | | (1-3 mo) |
| Focal-generalized-minor motor | | |
| Gelastic seizures | | (6-18 mo) |
| Motor control: | | |
| Level II: | Myoclonus-tremor | (2-4 mo) |
| Level I : | Hypotonia with loss of voluntary movements | (3-6 mo) |
| Level III + IV: | Hyperreflexia-opisthotonus dystonia (rigidity-spasticity) | (6-18 mo) |
| Intellectual function: | | |
| Progressive loss of early intellectual abilities | | (4-8 mo) |
| Dementia | | (2-3 yr) |
| Sensory perception: | | |
| Blindness | | (10-12 mo) |
| Deafness | | (∿18 mo) |

Table 3. CNS dysfunction in juvenile $G_{M2}$ gangliosidoses 1 - 10 years (variable onset) (Schulte 1984)

| | | |
|---|---|---|
| Excitability: | | |
| Seizures: (1-12 yrs) | | |
| Motor control: | | |
| Level II: | Gait disturbances, ataxia, nystagmus dysmetria intention tremor | (1-2 yrs) |
| Level IV: | Deterioration of speech pseudobulbar paralysis, mutism | (3-4 yrs) |
| Level IV: | Hyperreflexia clonus-Babinski spasticity | (2-4 yrs) |
| Level III: | Choreiform and athetoid movements decerebrate rigidity | (4-10 yrs) |
| Intellectual function: | | |
| Dementia | | (3-4 yrs) |
| Sensory preception: | | |
| Blindness (late in course of the disorder) | | |

Table 4. Variations in phenotype have been found for adult $G_{M2}$ gangliosidosis

| Clinical Diagnosis | Number of cases | Enzyme defect | References |
|---|---|---|---|
| Atypical spinocerebellar degeneration | 2 | Hex A | Rapin et al. (1976) |
| Spinocerebellar degeneration | 2 | Hex A + B | Oonk et al. (1979) |
| Amyothrophic lateral sclerosis like syndrome | 1 | Hex A | Yaffee et al. (1979) |
| Atypical Friedreich ataxia | 9 | Hex A | Willner et al. (1981) |
| Juvenile spinal muscular atrophy | 1 | Hex A | Johnson et al. (1982) |
| Muscle weakness stuttering speech | 4 | Hex A | Navon et al. (1981) |

## Degree of Enzyme Deficiency and Development of Disease

From Tables 2 - 4 a great clinical heterogeneity is obvious, though all types of variant B of $G_{M2}$ gangliosidosis are caused by a mutation of the α-subunit of hexosaminidase A and a neuronal storage of ganglioside $G_{M2}$. This heterogeneity is paralleled by the extent and pattern of glycolipid accumulation in different brain regions. Whereas the infantile type is characterized by an excessive and ubiquitous neuronal glycolipid accumulation (up to 12% of the brain dry weight), late onset forms show much less severe accumulation, which is restricted to specific regions; whereas the cortex is almost unimpaired, the hippocampus and nuclei of the brainstem, and spinal cord as well as the granular cells in the cerebellum and retina are mostly affected (Escola 1981, Terry and Weiss 1963, Jatzkewitz et al. 1965).

Thus different allelic mutations in one gene locus may lead to extremely variable clinical and neuropathological forms of the same biochemical variant of gangliosidosis. A crucial difference observed between the different clinical courses, e.g., infantile and adult forms, are different residual activities of the mutated hexosaminidase A against its natural substrate. For a complete understanding of the pathogenesis of the gangliosidoses, it would be very important to determine exactly the residual activity of a mutated enzyme against all possible natural substrates in vivo. Mutated enzymes might have altered substrate specificities, changed stabilities against proteolytic degradation or against thermal inactivation, and might be inhibited differently compared to their normal counterparts by components of the lysosomal environment, such as mucopolysaccharides.

Unfortunately, the in vivo activity of mutated enzymes cannot be determined directly. The experimental approaches that are so far considered to yield the best approximations and which seem to be the most suitable methods for pre- and postnatal diagnosis of these diseases are (a) feeding of natural substrate to cultivated fibroblasts and (b) in vitro determination of enzyme activity in fibroblast homogenates, with the natural substrate in the presence of the respective activator

Table 5. Degradation of ganglioside $G_{M2}$ fibroblast extracts. (Conzelmann et al. 1983)

| Probands | Ganglioside $G_{M2}$-degradation [pmol/h mg AU] Mean | Range |
|---|---|---|
| Normal | 535 | 296 - 762; n = 9 |
| Heterozygotes (different genotypes) | 285 | 121 - 395; n = 4 |
| Var. B | | |
| Infantile | 2.4 | 0.8 - 3.8; n = 5 |
| Juvenile | 15.8 | 13.6 - 18.0; n = 5 |
| Adult | 19.1 | 13.1 - 32.8; n = 9 |
| Var. O | | |
| Infantile | 6.6 | 3.6 - 9.5; n = 2 |
| Juvenile | 23.6 | 9.5 - 39.4; n = 3 |
| Healthy | 105 | 75 - 143; n = 2 |

protein. Data obtained with the latter method indicate some residual activity in adult patients of variant B, which is small but significantly higher than that of infantile forms of $G_{M2}$-gangliosidosis (Table 5) (Conzelmann et al. 1983). Residual activities in the range of 10 - 20% of mean control value already appear to be compatible with normal life. These findings stress the importance of small variations of the low residual enzyme activities, that are found in the patients, for the development of different clinical courses of a disease.

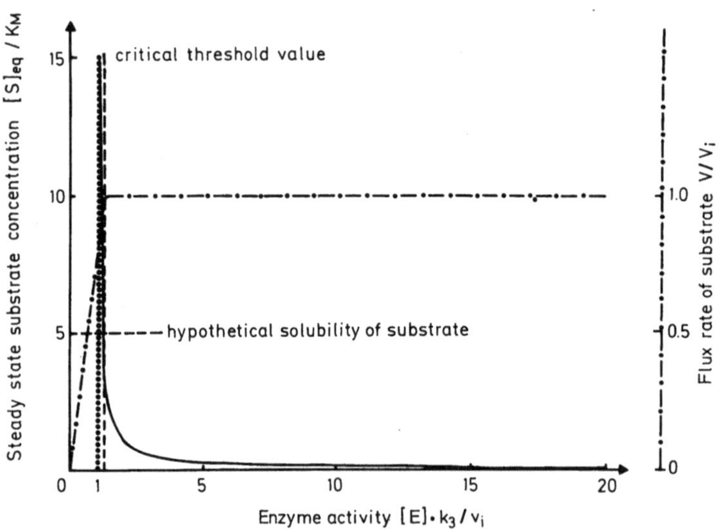


Fig. 4. Steady-state substrate concentration as a function of enzyme concentration and activity. (Conzelmann and Sandhoff 1983/84). The model underlying this theoretical calculation assumes influx of the substrate into a compartment at a constant rate ($v_i$) and its subsequent utilization by the enzyme (for details, see text). —— = $[S]_{eq}$, steady-state substrate concentration; ... = theoretical threshold of enzyme activity; --- = critical threshold value, taking limited solubility of substrate into account; -·-·- = turnover rate of substrate (flux rate)

Observations that moderate residual activities (e.g., 10% of normal) may still suffice to sustain normal catabolism of a substrate and that small variations within the low residual activities observed in patients (below 5% of normal) greatly influence the clinical course of the disease can easily be understood on the basis of a greatly simplified kinetic model (Fig. 4) (Conzelmann and Sandhoff 1983/84). Assuming a constant influx rate of the substrate, e.g., into the lysosomal compartment of an individual cell, and a degradation rate proportional to the degree of saturation of the available enzyme, the steady-state substrate concentration can be calculated as a function of the residual enzyme activity. For lysosomal enzymes, which are probably not regulated, this steady state concentration should be far below the $K_m$ value. It is then evident that even substantial reductions of the amount or turnover number of the enzyme will only lead to a moderate increase in the steady-state concentration of the substrate, but not in an accumulation, since the remaining activity is still sufficient to cope with the substrate's influx rate. Only when the residual activity falls below a critical threshold, will the overall turnover rate be more or less reduced. The residual turnover rate, and hence the rate at which the substrate accumulates, will then depend on the difference between threshold activity and actual residual enzyme activity in the lysosome. Since influx rate of the substrate as well as enzyme activity are different in different organs and cell types and may even vary between individual cells (e.g., neurons), the consequence of a mutation on different organs and cell types may be quite variable.

These calculations were performed as a first-order approximation on the basis of simple Michaelis-Menten kinetics without considering any complications which may arise, e.g., from regulatory properties of the enzyme considered or from the interaction of accumulating substrates with other components within the lysosomes. Still, they give an impression of how critical small variations of low residual enzyme activities (e.g., between 1% and 5% of normal) may be for the genesis or avoidance or for the severity of a lysosomal storage disorder.

## Possible Pathogenetic Factors Involved in the Development of Storage Disease

The pathogenic mechanisms involved in the development of neuronal ganglioside storage diseases are still unknown. Since the accumulating amphiphilic lipids are rather poorly soluble in aqueous solutions, they will precipitate within the cells in which they are synthesized, mainly within the lysosomal compartment. Although the storage compounds are normal, nontoxic components of the cell, their excessive accumulation in infantile cases (up to 12% of brain, dry weight) may mechanically destroy the neurons. However, in late onset forms the level of lipid accumulation is much lower than in infantile forms, therefore other mechanisms should also be discussed for the pathogenesis of gangliosidoses, e.g., an increase of ganglioside concentration in other membranes, thus disturbing the function of normal membrane-associated processes. Purpura and Suzuki (1976) demonstrated meganeurites and a tremendous increase of synaptic spines on nerve cells in postmortem brains of patients afflicted with ganglioside storage disease.

Another conceivable mechanism might be the formation of toxic compounds, as it occurs in Gaucher and Krabbe's disease. In these diseases, the occurrence of glucosylsphingosine and galactosylsphingosine has been shown (Svennerholm et al. 1980, Nilsson et al. 1982). These psychosine derivatives probably have lytic properties and are toxic in cell culture. Their progressive accumulation in brain tissue has been nicely demonstrated by Igisu and Suzuki (1984) in an animal model for Krabbe's disease, in the Twitcher mouse. The recent demonstration of a lysoganglioside $G_{M2}$ (gangliotriaosylsphingosine) accumulation (15 nmol/g wet weight) in a Tay Sachs-brain (Neuenhofer et al. 1986) may prove to be a starting point for the investigation of pathogenetic mechanisms in gangliosidoses.

Two further mechanisms might be considered. As demonstrated by Kint (1973), accumulating mucopolysaccharides inhibit ganglioside $G_{M1}$-β-galactosidase and thereby cause a decrease of its activity. In a similar way, accumulating glycolipids might inhibit enzymes of other pathways, causing a disturbance of the cellular metabolism. Finally, the tremendous accumulation observed in ballooned neurons of patients with infantile gangliosidoses may exhaust precursor pools for the biosynthesis of cellular components, e.g., pools of sphingosine, leading to an imbalance in metabolism.

It is to be expected that ganglioside metabolism is different in different types of neurons. Indeed, studies of the developmental profiles of individual gangliosides clearly show different ganglioside patterns for different brain regions (Vanier et al. 1971, Seyfried et al. 1983, Rösner 1982). Therefore it must be assumed that individual neurons will be affected differently by the same degree of deficiency in a catabolizing enzyme. More detailed knowledge on the glycoplipid metabolism of the various brain regions and its regulation is needed in order to formulate these questions more precisely.

## References


Banerjee A, Burg J, Conzelmann E, Carroll M, Sandhoff K (1984) Enzyme-linked immunosorbent assay for the ganglioside $G_{M2}$-activator protein - Screening of normal human tissues and body fluids, of tissues of $G_{M2}$ gangliosidosis, and for its subcellular localization. Hoppe-Seyler's Z Physiol Chem 365:347-356

Burg J, Conzelmann E, Sandhoff K, Solomon E, Swallow DM (1985) Mapping of the gene coding for the human $G_{M2}$ activator protein to chromosome 5. Ann Hum Genet 49:41-45

Carroll M, Robinson D (1973) Immunological properties of N-acetyl-β-D-glucosaminidase of normal human liver and of $G_{M2}$ gangliosidosis liver. Biochem J 131:91-96

Conzelmann E, Sandhoff K (1978) AB variant of infantile $G_{M2}$-gangliosidosis: Deficiency of a factor necessary for stimulation of hexosaminidase A-catalyzed degradation of ganglioside $G_{M2}$ and glycolipid $G_{A2}$. Proc Natl Acad Sci USA 75:3979-3983

Conzelmann E, Sandhoff K (1979) Purification and characterization of an activator protein for the degradation of glycolipids $G_{M2}$ and $G_{A2}$ by hexosaminidase A. Hoppe-Seyler's Z Physiol Chem 360:1837-1849

Conzelmann E, Sandhoff K (1983/84) Partial enzyme deficiencies: residual activities and the development of neurological disorders. Dev Neurosci 6:58-71

Conzelmann E, Sandhoff K, Nehrkorn H, Geiger B, Arnon A (1978) Purification, biochemical and immunological characterization of hexosaminidase A from variant AB of infantile $G_{M2}$-gangliosidosis. Eur J Biochem 84:27-33

Conzelmann E, Burg J, Stephan G, Sandhoff K (1982) Complexing of glycolipids and their transfer between membranes by the activator protein for lysosomal ganglioside $G_{M2}$ degradation. Eur J Bichem 123:455-464

Conzelmann E, Kytzia HJ, Navon R, Sandhoff K (1983) Ganglioside $G_{M2}$ N-acetyl-β-D-galactosaminidase activity in cultured fibroblasts of late-infantile and adult $G_{M2}$ gangliosidosis patients and of healthy probands with low hexosaminidase levels. Am J Hum Genet 35:900-913

D'Azzo A, Hoogeveen AT, Reuser AJJ, Robinson D, Galjaard H (1982) Molecular defect in combined β-galactosidase and neuraminidase deficiency in man. Proc Natl Acad Sci USA 79:4535-4539

Di Gregorio F, Ferrari G, Marine P, Siliprandi R, Gorio A (1984) The influence of gangliosides on neurite growth and regeneration. Neuropediatrics 15:93-96

Erickson AH, Blobel G (1979) Early events in the biosynthesis of the lysosomal enzyme cathepsin D. J Biol Chem 254:11771-11774

Escola J (1961) Über die Prozeßausbreitung der amaurotischen Idiotie im Zentralnervensystem in verschiedenen Lebensaltern und Besonderheiten der Spätform gegenüber der Pigmentatrophie. Arch Psychiat Nervenkr 202:95-112

Gautron S, Poenaru L, Boué J, Puissant H, Lisman JJW, Dreyfus J-C (1983) Evidence for the presence of β-subunit of hexosaminidase in a case of Sandhoff disease using a blotting technique. Hum Genet 63:258-261

Geiger B, Arnon R, Sandhoff K (1977) Immunochemical and biochemical investigation of hexosaminidase S. Am J Hum Genet 29:508-522

Hakomori SI (1975) Structures and organization of cell surface glycoplipids, dependency of cell growth and malignant transformation. Biochim Biophys Acta 417:55-89

Hasilik A (1980) Biosynthesis of lysosomal enzymes. Trends Biochem Sci 5:237-240

Hasilik A, Neufeld EF (1980) Biosynthesis of lysosomal enzymes in fibroblasts. Synthesis as precursors of higher molecular weight. J Biol Chem 255:4937-4945

Hasilik A, von Figura K, Conzelmann E, Nehrkorn H, Sandhoff K (1982) Lysosomal enzyme precursors in human fibroblasts. Activation of cathepsin D-precursors in vitro and activity of β-hexosaminidase A precursor towards ganglioside $G_{M2}$. Eur J Biochem 125:317-321

Hers HG (1965) Inborn lysosomal diseases. Gastroenterology 48:625-633

Igisu H, Suzuki K (1984) Progressive accumulation of a toxic metabolite in a genetic leukodystrophy. Science 224:753-755

Jatzkewitz H, Pilz H, Sandhoff K (1965) Quantitative Bestimmungen von Gangliosiden und ihren neuraminsäurefreien Derivaten bei infantilen, juvenilen und adulten Formen der amaurotischen Idiotie und einer spätinfantilen biochemischen Sonderform. J Neurochem 12:135-144

Johnson WG, Wigger HJ, Karp HR, Glaubiger LM, Rowland LP (1982) Juvenile spinal muscular atrophy, a new hexosaminidase deficiency phenotype. Ann Neurol 11:11-16

Kint JA (1973) Antagonistic action of chondroitin sulphate and cetylpyridinium chloride on human liver β-galactosidase. FEBS Lett 36:53-56

Kornfeld S, Reitman M, Varki A (1983) Role of carbohydrates in targeting newly synthesized lysosomal enzymes to lysosomes. In: Popper H, Reutter W, Gudat F, Köttgen E (eds) Structural carbohydrates in the liver. MTP Press, Boston, pp 175-180

Kresse H, Fuchs W, Glössl J, Holtfrerich D, Gilberg W (1981) Liberation of N-acetylglucosamine-6-sulfate by human N-acetylhexosaminidase A. J Biol Chem 256:12926-12932

Kytzia HJ, Sandhoff K (1985) Evidence for two different active sites on human hexosaminidase A - Interaction of $G_{M2}$ activator protein with hexosaminidase A. J Biol Chem 260:7568-7572

Kytzia HJ, Hinrichs U, Maire I, Suzuki K, Sandhoff K (1983) Variant of $G_{M2}$ gangliosidosis with hexosaminidase A having a severely changed substrate specificity. EMBO J 2:1201-1205

Li SC, Hirabayashi Y, Li YT (1981) A new variant of type AB $G_{M2}$-gangliosidosis. Biochem Biophys Res Commun 101:479-485

Markwell MAK, Svennerholm L, Paulson JC (1981) Specific gangliosides function as host cell receptors for Sendai virus. Proc Natl Acad Sci USA 78:5406-5410

Navon R, Argov Z, Brandt N, Sandbank U (1981) Adult $G_{M2}$ gangliosidoses in association with Tay-Sachs disease: a new phenotype. Neurology (Minneap) 31:1397-1401

Navon R, Kopel R, Nutman J, Frisch A, Conzelmann E, Sandhoff K, Adam A (1985) Hereditary heat-labile hexoaminidase B: A variant whose homozygotes synthesize a functional hex A. Am J Hum Genet 37:138-146

Neuenhofer S, Sandhoff K (1985) Affinity labeling of the $G_{M2}$-activator protein. FEBS Lett 185:112-114

Neuenhofer S, Conzelmann E, Schwarzmann G, Egge H, Sandhoff K (1986) Occurrence of Lysoganglioside Lyso-$G_{M2}$ ($II^3$-$Neu^5Ac$-Gangliotriaosylsphingosine) in $G_{M2}$ Gangliosidosis Brain. Biol Chem Hoppe-Seyler, in press

Neufeld EF, D'Azzo A, Proia RL (1984) Defective synthesis or maturation of the α-chain of β-hexosaminidase in classic and variant forms of Tay-Sachs disease. In: Barranger JA, Brady RO (eds) Molecular basis of lysosomal storage disorders. Academic Press, London New York, pp 251-256

Nilsson O, Månsson JE, Håkansson G, Svennerholm L (1982) The occurrence of psychosine and other glycolipids in spleen and liver from the three major types of Gaucher's disease. Biochim Biophys Acta 712:453-463

O'Brien JS (1983) The gangliosidoses. In: Stanbury JS, Wyngaarden JB, Fredrickson DS, Goldstein JL, Brown MS (eds) The metabolic basis of inherited disease, 5th edn. McGraw Hill, New York, pp 945-969

Oonk JGW, van der Helm HJ, Martin JJ (1979) Spinocerebellar degeneration: Hexosaminidase A and B deficiency in two adult sisters. Neurology 29:380-384

Proia RL, Neufeld EF (1982) Synthesis of β-hexosaminidase in cell-free translation and in intact fibroblasts: An insoluble precursor α-chain in a rare form of Tay-Sachs disease. Proc Natl Acad Sci USA 79:6360-6364

Proia RL, D'Azzo A, Neufeld EF (1984) Association of α- and β-subunits during the biosynthesis of β-hexosaminidase in cultured human fibroblasts. J Biol Chem 259: 3350-3354

Purpura DP, Suzuki K (1976) Distortion of neuronal geometry and formation of aberrant synapses in neuronal storage disease. Brain Res 116:1-21

Rapin I, Suzuki K, Suzuki K, Valsamis MP (1976) Adult (chronic) $G_{M2}$ gangliosidosis. Arch Neurol 33:120-130

Rösner H (1982) Ganglioside changes in the chicken optic lobes as biochemical indicators of brain development and maturation. Brain Res 236:49-61

Sandhoff K (1977) The biochemistry of sphingolipid storage diseases. Angew Chem Int Ed Engl 16:273-285

Sandhoff K (1984a) Biosynthese und Abbau von Glykosphingolipiden, Stoffwechsel an Phasengrenzflächen. Funkt Biol Med 3:141-151

Sandhoff K (1984b) Function and relevance of activator proteins for glycolipid degradation. In: Barranger JA, Brady RO (eds) Molecular basis of lysosomal storage disorders. Academic Press, London New York, pp 19-49

Sandhoff K, Christomanou H (1979) Biochemistry and genetics of gangliosidoses. Hum Genet 50:107-143

Sandhoff K, Conzelmann E (1984) The biochemical basis of gangliosidoses. Neuropediatrics Suppl 15:85-92

Sandhoff K, Andreae U, Jatzkewitz H (1968) Deficient hexosaminidase activity in an exceptional case of Tay-Sachs disease with additional storage of kidney globoside in visceral organs. Pathol Eur 3:278-285

Sandhoff K, Harzer K, Wässle W, Jatzkewitz H (1971) Enzyme alterations and lipid storage in three variants of Tay-Sachs disease. J Neurochem 18:2469-2489

Schachter H, Roseman S (1980) Mammalian glycosyltransferases: their role in the synthesis and function of complex carbohydrates and glycolipids. In: Lennarz WJ (ed) The biochemistry of glycoproteins and proteoglycans, pp 85-160

Schulte FJ (1984) Clinical course of $G_{M2}$ gangliosidoses, A correlative attempt. Neuropediatrics Suppl 15:66-70

Seyfried TN, Miyazawa N, Yu RK (1983) Cellular localization of gangliosides in the developing mouse cerebellum: analysis using the weaver mutant. J Neurochem 41: 491-505

Sly WS, Fischer HD (1982) The phosphomannosyl recognition system for intracellular and intercellular transport of lysosomal enzymes. J Cell Biochem 18:67-85

Sonderfeld S, Conzelmann E, Schwarzmann G, Burg J, Hinrichs U, Sandhoff K (1985) Incorporation and metabolism of ganglioside $G_{M2}$ in normal and $G_{M2}$ gangliosidosis skin fibroblasts. Eur J Biochem 149:247-255

Stanbury JB, Wyngaarden JB, Fredrickson DS, Goldstein JL, Brown MS (1983) The metabolic basis of inherited disease, 5th edn. McGraw Hill, New York

Stevens RL, Fluharty AL, Kihara H, Kaback MM, Shapiro LJ, Marsh B, Sandhoff K, Fischer G (1981) Cerebroside sulfatase activator deficiency induced Metachromatic Leukodystrophy. Am J Hum Genet 33:900-906

Svennerholm L, Vanier MT, Månsson J-E (1980) Krabbe disease: a galactosylphingosine (psychosine) lipidosis. J Lipid Res 21:53-64

Terry RD, Weiss M (1963) Studies in Tay Sachs disease II - ultrastructure of the cerebrum. J Neuropathol Neurol 22:18-55

Vanier MT, Holm M, Öhmann R, Svennerholm L (1971) Developmental profiles of gangliosides in human and rat brain. J Neurochem 18:581-592

Von Figura K, Hasilik A (1984) Genesis of lysosomal enzyme deficiencies. Trends Biochem Sci 9:29-31

Von Figura K, Steckel F, Hasilik A (1983) Juvenile and adult metachromatic leukodystrophy. Partial restoration of arylsulfatase A (cerebroside sulfatase) activity by inhibitors of thiol proteinases. Proc Natl Acad Sci USA 80:6066-6070

Willner JP, Grabowski GA, Gordon RE, Bender AN, Desnick RJ (1981) Chronic $G_{M2}$ gangliosidosis masquerading as atypical Friedreich ataxia clinical morphological and biochemical studies of nine cases. Neurology 31:787-798

Yaffe MG, Kaback MM, Goldberg M, Miles J, Itabashi H, McIntyre H, Mohandas T (1979) An amyotrophic lateral sclerosis-like syndrome with hexosaminidase A deficiency: a new type of $G_{M2}$-gangliosidosis. Neurology 29:611-611

Yamakawa T, Nagai Y (1978) Glycolipids at the cell surface and their biological functions. Trends Biochem Sci 3:128-131

# The Apamin-Sensitive $Ca^{2+}$-Dependent $K^+$ Channel. Molecular Properties, Differentiation, and Endogeneous Ligands in Mammalian Brain

M. Lazdunski, M. Fosset, M. Hugues, C. Mourre, J. F. Renaud, G. Romey, and H. Schmid-Antomarchi[1]

## Apamin, Its Structure and Its Active Site

Apamin is a neurotoxin extracted from bee venom (Habermann 1972). It is a polypeptide of 18 amino acids with two disulfide bridges. It is the only polypeptide neurotoxin, as far as we know, that passes the blood-brain barrier. Analysis of the structure-function relationships of this toxin has shown that two of the 18 amino acids in the sequence have particular importance for the action of the toxin, they are $Arg_{13}$ and $Arg_{14}$ (Vincent et al. 1975). These two residues seem to be essential elements of the active site of the toxin. Chemical modifications elsewhere in the sequence may decrease the toxicity of the polypeptide but do not suppress its biological activity. Cumulative chemical modifications – of the amino group and of the imidazole of $His_{18}$ for example – may, however, abolish the activity of the toxin (Vincent et al. 1975). Solid phase synthesis of apamin and analogs has been carried out (Cosand and Merrifield 1977, Granier et al. 1978). This approach has confirmed that the active site of apamin comprises the two residues $Arg_{13}$ and $Arg_{14}$.

The exact three-dimensional structure of the toxin remains unknown. However, recent solution analysis of apamin using NMR techniques has suggested that the toxin is highly ordered with an α-helical core and regions of β-type turns (Bystrov et al. 1980, Wemmer and Kallenbach 1983).

## Apamin Blocks $Ca^{2+}$-Dependent $K^+$ Channels

Apamin does not seem to interact with receptors of the most classical neurotransmitters (Vincent et al. 1975). The toxin became a good candidate as a specific blocker of a $Ca^{2+}$-dependent $K^+$ conductance when $K^+$ flux studies showed that apamin prevented the rise in potassium permeability in guinea pig hepatocytes and taenia ceci (Banks et al. 1979, Maas et al. 1980) produced by ATP and noradrenaline.

A direct demonstration of the specific action of the toxin was obtained using voltage-clamp techniques with both neuroblastoma cells and rat muscle cells in culture (Hugues et al. 1982a,b). The bee venom toxin suppresses the after-hyperpolarization potential (a.h.p.) and it was shown by voltage-clamp that this blockade is due to an inhibition of the $Ca^{2+}$-dependent $K^+$ conductance.

---

[1]Centre de Biochimie du CNRS, Faculté des Sciences, Parc Valrose, F-06034 Nice Cedex, France

36. Colloquium - Mosbach 1985
Neurobiochemistry

## Biochemical Properties of the Apamin Binding Component of the $Ca^{2+}$-Dependent $K^+$ Conductance

Radiolabeled monoiodoapamin can be prepared at a very high specific radioactivity (± 2000 Ci/mmol) by incorporating iodine on $His_{18}$, the C-terminal residue of the toxin (Hugues et al. 1982c, Habermann and Fischer 1979). The mean lethal dose ($LD_{50}$) of the monoiodo derivative of apamin measured by intracisternal injection in mice is 4.2 ± 1 µg/kg as compared to 2 ± 0.5 µg/kg for the native toxin (Hugues et al. 1982c). This decrease of effectiveness by a factor of about 2 after iodination is also observed using a bioassay in which apamin inhibited (and even converted to a contraction) the neurotensin-induced relaxation in segments of guinea pig proximal colon (Hugues et al. 1982c).

The main properties of the association of $^{125}I$-apamin binding to the $Ca^{2+}$-dependent $K^+$ channel are the following: (1) the specific binding component is much higher than the nonspecific binding component, (2) there is only one family of binding sites (the Scatchard plot is linear), (3) the affinity of apamin for its receptor is high; it is 15 - 25 pM for the monoiodo derivative and 10 pM for the native toxin, (4) the binding of apamin to synaptosomes is characterized by a very low binding capacity; synaptosomes only bind 12 - 13 fmol of $^{125}I$-apamin/mg of protein and therefore contain 150 to 300 times less apamin-sensitive channels than tetrodotoxin-sensitive $Na^+$ channels (Hugues et al. 1982c).

$^{125}I$-apamin has now been used to identify apamin-sensitive $Ca^{2+}$-dependent $K^+$ channels in a variety of cell types including neuroblastoma cells (Hugues et al. 1982a), smooth muscle (Hugues et al. 1982d), skeletal muscle cells in culture (Hugues et al. 1982b), and hepatocytes (Cook et al. 1983). The high affinity of the toxin for its receptor and the low number of binding sites have been found in all systems. In neuroblastoma or muscle cells in culture, the number of apamin-sensitive $Ca^{2+}$-dependent $K^+$ channels is 5 - 7 times lower than the number of tetrodotoxin-sensitive $Na^+$ channels (Hugues et al. 1982a,b).

The other properties of the apamin-receptor complex which are of interest are the following ones: (1) the toxin dissociates very slowly from its receptor. Dissociation rate constants of $1.5 - 4 \times 10^{-4}$ $s^{-1}$ have been found in the different nerve and muscle system investigated (Hugues et al. 1982a,c); they correspond to half-lifes of dissociations which can be as low as about 60 min; (2) nontoxic derivatives of apamin do not bind to the apamin receptor; (3) the binding of monoiodoapamin to its receptor is sensitive to cations (Hugues et al. 1982c; Habermann and Fischer 1979). $K^+$ and $Rb^+$ at concentrations between 10 µM and 5 mM are able to increase the binding of $^{125}I$-apamin to its receptor by a factor of about 2 while each of the many other cations tested ($Li^+$, $Na^+$, guanidinium, etc.) completely inhibits $^{125}I$-apamin binding in the concentration range of 1 - 100 mM. These results have been interpreted (Hugues et al. 1982c) as suggesting the existence of two different binding sites for cations. In this hypothesis, site 1 is specific for $K^+$ and $Rb^+$ and binds these cations with an affinity corresponding to a dissociation constant of about 500 µM. This site is distinct from the binding site of apamin; its occupancy by $K^+$ or $Rb^+$ results in an increased affinity of the receptor for the toxin. Site 2 most probably belongs to the apamin binding site itself; it recognizes every cation tested with a low affinity. This anionic site which serves as a cation binding site is probably the site to which $Arg_{13}$ and $Arg_{14}$, the two crucial residues of the toxin, bind. It is then not very surprising that molecules which contain guanidinium groups like guanidinium itself, amiloride, or neurotensin (which has two contiguous arginine residues like apamin) possess a higher affinity for the apamin receptor than other inorganic cations or positively charged molecules (Hugues et al. 1982a,c). Among this last class of molecules, quinine and quinidine prevent $^{125}I$-apamin binding to the $Ca^{2+}$-dependent $K^+$ channel with $K_{0.5}$ values of 100 - 200 µM (Hugues et al. 1982b).

## Apamin as a Tool to Purify the Apamin-Sensitive $Ca^{2+}$-Dependent $K^+$ Channel and to Determine Its Molecular Weight and Its Polypeptide Composition

The apamin receptor is of a polypeptide nature (Hugues et al. 1982a,c) like the receptor of the many neurotoxins which are known to be specific for the fast $Na^+$ channel (Lazdunski and Renaud 1982). The problem is to know whether apamin can be used to purify the $Ca^{2+}$-dependent $K^+$ channel like tetrodotoxin, saxitoxin or scorpion toxins have been used for the purification of the $Na^+$ channel (Barchi et al. 1980, Hartshorne and Catterall 1981, Moore et al. 1982, Barhanin et al. 1983a, Norman et al. 1983a). Properties of apamin in favor of its successful use to purify the $Ca^{2+}$-dependent $K^+$ channels are its high affinity for its receptor and the slow rate of dissociation of the complex. However, a clear difficulty for visualizing this purification in an optimistic way is due to the very small number of $Ca^{2+}$-dependent $K^+$ channels in all preparations which have been assayed up till now. The tetrodotoxin-sensitive $Na^+$ channel has been purified from brain synaptosomes (Hartshorne and Catterall 1981, Barnhanin et al. 1983a) but, as it has been seen before, $Na^+$ channels are 150 - 300 times more numerous in this preparation than apamin-sensitive $Ca^{2+}$-dependent $K^+$ channels and brain synaptosomes are unfortunately the best source of apamin receptor presently identified.

Available data concerning the structure of the apamin-sensitive channel have been obtained directly on membranes. The molecular weight ($M_r$) of the apamin receptor was determined using the radiation inactivation technique which was so successful in establishing the $M_r$ of the $Na^+$ channel (Levinson and Ellory 1973, Barhanin et al. 1983b).

Affinity labeling of the apamin-sensitive $Ca^{2+}$-dependent $K^+$ channel was realized successfully by cross-linking the toxin to its receptor on the channel structure using disuccinimidyl suberate (DSS). To increase changes of success, the experiment was carried out on synaptic membranes which contain about twice as much apamin-sensitive $Ca^{2+}$ channel as synaptosomes, and at pH 9 where the binding capacity receptor is maximum. The covalent labeling indicates that the apamin receptor in the $Ca^{2+}$-dependent $K^+$ channel of synaptic membranes is a single polypeptide chain of a $M_r$ of about 30.0000 (Hugues et al. 1982e, Schmid-Antomarchi et al. 1984).

It is not known at the present time whether the $Ca^{2+}$-dependent $K^+$ channel is made of only one type of polypeptide chain ($M_r$ = 30.000) or whether there are other polypeptide chains which have not been labeled by apamin. If there is only one type of polypeptide chain the channel is then an oligomeric structure containing 8 ± 2 chains of $M_r$ 30.000.

Knowing that the $M_r$ of the apamin receptor is near 250.000, one can easily calculate that the specific activity of the pure preparation of apamin-sensitive $Ca^{2+}$-dependent $K^+$ channel will have to be near 4000 pmol/mg of protein. The amount of apamin receptor in synaptosomes being about 12 fmol/mg of protein (Hugues et al. 1982c), the isolation of the pure channel will require a purification by a factor of 300,000 - 400,000. We have successfully solubilized the apamin receptor with a complete preservation of the same binding properties it had in the membrane. However, while classical steps of purification have been sufficient to purify the tetrodotoxin-sensitive $Na^+$ channel (Hartshorne and Catterall 1981, Barhanin et al. 1983a), it will certainly be necessary to use affinity columns containing either apamin or anti-$Ca^{2+}$-dependent $K^+$ channel antibodies to successfully isolate this channel in the pure form.

## The Apamin-Sensitive $Ca^{2+}$-Dependent $K^+$ Channel is Only One of Several Types of $Ca^{2+}$-Dependent $K^+$ Channels

Since the development of patch-clamp techniques (Neher et al. 1978) single channel recordings from $Ca^{2+}$-dependent $K^+$ channels have been reported from various preparations such as bovine chromaffin cells (Marty 1981), rat myotubes (Barrett et al. 1982, Methfessel and Boheim 1982), bullfrog ganglion cells (Adams et al. 1982) and rabbit muscle T-tubule membrane fragments reconstituted into planar lipid bilayers (Latorre et al. 1982). In each case, the channel shows a large conductance (100 - 250 pS) and is highly selective for $K^+$ ions. One of the important problems to solve is to know whether the a.h.p. which is inhibited by apamin in several cellular systems is due to the same type of $Ca^{2+}$-dependent $K^+$ channel that has been identified so successfully using patch-clamp techniques. To solve this problem we have therefore carried out detailed work on rat skeletal muscle cells in culture. These cells have the following properties: (1) they have an a.h.p. which is inhibitable by apamin and which is due to a $Ca^{2+}$-dependent $K^+$ channel (Hugues et al. 1982b), (2) they have $^{125}I$-apamin receptors, (3) they have $Ca^{2+}$-dependent $K^+$ channels which can be identified by patch-clamp techniques (Barrett et al. 1982). Our studies have shown that $Ca^{2+}$-dependent $K^+$ channels with a large conductance which are identified by patch-clamp techniques are inhibitable by TEA while they are absolutely insensitive to apamin (Romey and Lazdunski 1984). Conversely $Ca^{2+}$-dependent $K^+$ channels, which generate the a.h.p., and which can be identified by voltage-clamp, are very sensitive to apamin and insensitive to TEA (Romey and Lazdunski 1984). These results, which have also been obtained with neuroblastoma cells, clearly indicate the existence of different types of electrically expressed $Ca^{2+}$-dependent $K^+$ channels with a different pharmacology and a different physiological function. The function of the apamin-sensitive $Ca^{2+}$-dependent $K^+$ channels is to generate a.h.p.'s; the function of the $Ca^{2+}$-dependent $K^+$ channels identified as large conductance channels by patch-clamp techniques may be to prevent a prolonged depolarization of the cell.

In spite of extensive work, no successful recording of single channel activity from apamin-sensitive $Ca^{2+}$ channels has been obtained up till now. This may be due to the small number of these channels.

## Autoradiographic Localization of Apamin-Sensitive $Ca^{2+}$-Dependent $K^+$ Channels in Rat Brain

Autoradiograms of the apamin binding sites on $Ca^{2+}$-dependent $K^+$ channels in mammalian brain have been obtained (Mourre et al. 1984). More than 90% of $^{125}$I-apamin binding to rat brain sections was specific binding.

The apamin-sensitive $Ca^{2+}$-dependent $K^+$ channel is in many sections of the brain. High grain densities are present in cortex and mainly in the internal pyramidal layer of the cortex, in the olfactory nucleus, in the lateral septal nucleus, in dentate gyrus, Ammon's horn, subiculum, habenula, geniculate nucleus, anterior thalamus, colchear nucleus, nucleus of spinal tract of the trigeminal nerve, inferior olive, gracilus and cuneate nuclei and in the granular layer of the cerebellum. Moderate amounts of $^{125}$I-apamin binding sites were also found in colliculi, caudate putamen, and in vestibular and red nuclei. The central gray, hypothalamus, substantia nigra and cervical nuclei were weakly labeled. The white matter was not labeled.

## Developmental Properties of the $Ca^{2+}$-Dependent $K^+$ Channel in Mammalian Skeletal Muscle and the All-or-None Role of Innvervation (Schmid-Antomarchi et al. 1985)

The long-lasting after hyper-polarization (a.h.p.) which follows the action potential in rat myotubes differentiated in culture is due to $Ca^{2+}$-activated $K^+$ channels. These channels have the property to be specifically blocked by the bee venom toxin apamin at low concentrations. Apamin has been used in this work to analyze by electrophysiological and biochemical techniques the role of innervation in the expression of these important channels. The main results are as follows: (1) Long-lasting a.h.p.'s which follow the action potential in rat myotubes in culture disappear when myotubes are co-cultured with nerve cells from the spinal cord under the conditions of in vitro innervation. (2) Extensor digitorum longus muscles from adult rats have action potentials which are not followed by a.h.p.'s but a.h.p.'s are systematically recorded after muscle denervation and they are blocked by apamin. (3) Specific $^{125}$I-apamin binding is undetectable in innervated muscle fibers but it becomes detectable 2 - 4 days after muscle denervation to be maximal 10 days after denervation. (4) Apamin receptors detected with $^{125}$I-apamin are present at foetal stages with biochemical characteristics identical to those found in myotubes in culture. The receptor number decreases as maturation proceeds and $^{125}$I-apamin receptors completely disappear after the first week of post-natal life in parallel with the disappearance of multi-innervation. All these results taken together strongly suggest an all-or-none effect of innervation on the expression of apamin-sensitive $Ca^{2+}$-activated $K^+$ channels.

## An Endogeneous Apamin-Like Factor Modulating $Ca^{2+}$-Dependent $K^+$ Channels Activity Exists in Mammalian Brain (Fosset et al. 1984)

An apamin-like factor has been purified from pig brain after acidic extraction of the tissue and several steps of purification on sulfopropyl Sephadex C25 and C18 reverse phase high pressure liquid chromatography (Fosset et al. 1984). The apamin-like activity was followed during the purification procedure using two biochemical assays: a radio receptor assay (Hugues et al. 1982c) and a radioimmunoassay using anti-apamin antibodies (Schweitz and Lazdunski 1984) and two physiological assays:

the measure of the contractile activity of a guinea pig taenia coli (Hugues et al. 1982d), and the electrophysiological measurement of the ability of fractions to block the a.h.p. in rat skeletal muscle cells (Hugues et al. 1982b).

A purification of an apamin-like factor has been achieved. Its properties are the following: (1) The factor is able to prevent $^{125}$I-apamin to its binding site on rat brain synaptosomes. (2) The factor antagonizes $^{125}$I-apamin recognition by anti-apamin antibodies. (3) The factor contracts the intestinal smooth muscle previously released by epinephrine. (4) The factor selectively blocks the hyperpolarization that follows its action potential on rat skeletal myotube in culture. (4) The factor is sensitive to the action of trypsin and insensitive to chymotrypsin digestion.

All these properties are those of apamin itself. These results strongly suggest the presence in pig brain of an endogeneous equivalent of apamin. Knowing that a purification procedure gives an activity equivalent to 1.5 ± 0.5 pmol of apamin per pig brain, large scale purification will be necessary to obtain enough apamin-like factor to sequence it. Our results suggest that inactive precursors of the apamin-like factor could also be present in the pig brain, the structure of the native precursor will be known once the cloning of the gene encoding for this precursor will be achieved.

## References


Adams PR, Constanti A, Brown DA, Clark RB (1982) Fast voltage-sensitive potassium current in vertebrates sympathetics neurons. Nature (London) 296:746-749

Banks BEC, Brown C, Burgess GM, Burnstock G, Claret M, Cocks TM, Jenkinson DH (1979) Apamin blocks certain neurotransmitter-induced increases in potassium permeability. Nature (London) 282:415-417

Barchi RL, Cohen SA, Murphy LE (1980) Purification from rat sarcolemma of the saxitoxin binding component of the excitable membrane sodium channel. Proc Natl Acad Sci USA 77:1306-1310

Barhanin J, Pauron D, Lombet A, Norman RI, Vijverberg HPM, Giglio JR, Lazdunski M (1983a) Electrophysiological characterization, solubilization and purification of the *Tityus* γ toxin receptor associated with the gating component of the $Na^+$ channel from rat brain. EMBO J 2:915-920

Barhanin J, Schmid A, Lombet A, Wheeler KP, Lazdunski M (1983b) Molecular size of different neurotoxin receptors on the voltage-sensitive $Na^+$ channel. J Biol Chem 258:700-702

Barrett JN, Magleby KL, Pallotta BS (1982) Properties of single calcium-activated potassium channels in cultured rat muscle. J Physiol (London) 331:211-230

Bystrov VF, Okhanov VV, Miroshnikov AI, Ovchinnikov YA (1980) Solution spatial structure of apamin as derived from NMR study. FEBS Lett 119:113-117

Cook NS, Haylett DN, Strong P (1983) High affinity binding of $^{125}$I-monoiodo apamin to isolated guinea-pig hepatocytes. FEBS Lett 152:265-269

Cosand WL, Merrifield RB (1977) Concept of internal structural controls for evaluation of inactive synthetic peptide analog: synthesis of (ORN 13, 14) apamin and its guanidination to an apamin derivative with full neurotoxic activity. Proc Natl Acad Sci USA 74:2771-2775

Fosset M, Schmid-Antomarchi H, Hugues M, Romey G, Lazdunski M (1984) The presence in pig brain of an endogeneous equivalent of apamin, the bee venom peptide which specifically blocks $Ca^{2+}$-dependent $K^+$ channels. Proc Natl Acad Sci USA 81:7228-7232

Granier C, Pedroso Muller E, Van Rietschoten J (1978) Use of synthetic analogs for a study on the structure-activity rleationship of apamin. Eur J Biochem 82:293-299

Habermann E (1972) Bee and wasp venom. Science 177:314-322

Habermann E, Fischer K (1979) Bee venom neurotoxin (apamin): iodine labeling and characterization of binding sites. Eur J Biochem 94:355-364
Hartshorne RP, Catterall WA (1981) Purification of the saxitoxin receptor of the sodium channel from rat brain. Proc Natl Acad Sci USA 78:4620-4624
Hugues M, Romey G, Duval D, Vincent JP, Lazdunski M (1982a) Apamin as a selective blocker of the calcium-dependent potassium channel in neuroblastoma cells: voltage-clamp and biochemical characterization of the toxin receptor. Proc Natl Acad Sci USA 79:1308-1312
Hugues M, Schmid H, Romey G, Duval D, Frelin C, Lazdunski M (1982b) The calcium-dependent slow potassium conductance in cultured rat muscle cells: characterization with apamin. EMBO J 9:1039-1042
Hugues M, Duval D, Kitabgi P, Lazdunski M, Vincent JP (1982c) Preparation of a pure monoiodo derivative of the bee venom neurotoxin apamin and its binding properties to rat brain synaptosomes. J Biol Chem 257:2762-2769
Hugues M, Duval D, Schmid H, Kitabgi P, Lazdunski M (1982d) Specific binding and pharmacological interactions of apamin, the neurotoxin from bee venom, with guinea-pig colon. Life Sci 31:437-443
Hugues M, Schmid H, Lazdunski M (1982e) Identification of a protein component of the calcium-dependent potassium channel by affinity labeling with apamin. Biochem Biophys Res Commun 107:1577-1582
Kepner GR, Macey RI (1968) Membrane enzyme systems. Molecular size determination by irradiation inactivation. Biochem Biophys Acta 163:188-203
Latorre R, Vergara C, Hidalgo C (1982) Reconstitution in planar lipid bilayers of a calcium-dependent potassium channel from transverse tubule membranes isolated from rabbit skeletal muscle. Proc Natl Acad Sci USA 79:805-809
Lazdunski M, Renaud JF (1982) The action of cardiotoxin on cardiac plasma membranes. Annu Rev Physiol 44:463-473
Levinson SR, Ellory JC (1973) Molecular size of the tetrodotoxin binding site estimated by irradiation inactivation. Nature (London) 245:122-123
Maas AD, Den Hertog A (1979) The effect of apamin on the smooth muscle cells of the guinea-pig taenia coli. Eur J Pharmacol 58:151-156
Maas ADJJ, Den Hertog A, Ras R, Van Den Akker J (1980) The action of apamin on guinea-pig taenia caeci. Eur J Pharmacol 67:265-274
Marty A (1981) Calcium-dependent potassium channels with large unitary conductance in chromaffin cell membranes. Nature (London) 291:497-500
Methfessel C, Boheim G (1982) The gating of single calcium-dependent potassium channel is described by an activation blockade mechanism. Biophys Struct Mech 9:35-60
Moore HPM, Fritz LC, Raftery MA, Brokes JP (1982) Isolation and characterization of a monoclonal antibody against the saxitoxin-binding component from the electric organ of the eel *Electrophorus electricus*. Proc Natl Acad Sci USA 79:1673-1677
Mourre C, Schmid-Antomarchi H, Hugues M, Lazdunski M (1984) Autoradiographic localization of apamin-sensitive $Ca^{2+}$-dependent $K^+$ channels in rat brain. Eur J Pharmacol 100:135-136
Neher E, Sakmann B, Steinbach JH (1978) The extracellular patch-clamp: a method for resolving current through individual open channels in biological membranes. Pflügers Archiv 375:219-228
Norman RI, Schmid A, Lombet A, Barhanin J, Lazdunski M (1983a) Purification of binding protein for *Tityus* γ toxin identified with the gating component of the voltage-sensitive $Na^+$ channel. Proc Natl Acad Sci USA 80:4164-4168
Norman RI, Borsotto M, Fosset M, Lazdunski M, Ellory JC (1983b) Determination of the molecular size of the nitrendipine-sensitive $Ca^{2+}$-channel by radiation inactivation. Biochem Biophys Res Commun 111:878-883
Romey G, Lazdunski M (1984) The coexistence in rat muscle cells of two distinct classes of $Ca^{2+}$-dependent $K^+$ channels with different pharmacological properties and different physiological functions. Biochem Biophys Res Commun 118:669-674
Schmid-Antomarchi H, Hugues M, Norman RI, Ellory JC, Borsotto M, Lazdunski M (1984) Molecular properties of the apamin-sensitive $Ca^{2+}$-dependent $K^+$ channel: radiation: inactivation, affinity labelling and solubilization. Eur J Biochem 142:1-6

Schmid-Antomarchi H, Renaud JF, Romey G, Hugues M, Schmid A, Lazdunski M (1985) The all-or-none role of innervation in the expression of the apamin-sensitive $Ca^{2+}$-activated $K^+$ channel in mammalian skeletal muscle. Proc Natl Acad Sci USA 82:2188-2195
Schweitz H, Lazdunski M (1984) A microradioimmunoassay for apamin. Toxicon 22: 985-988
Vincent JP, Schweitz H, Lazdunski M (1975) Structure-function relationships and site of action of apamin. A neurotoxic polypeptide of bee venom with an action on the central nervous system. Biochemistry 14:2521-2525
Vladimirova IA, Shuba MF (1978) Effect of strychnine, hydrastine and apamin on synaptic transmission in smooth muscle cells. Neurofiziologija 10:295-299
Wemmer D, Kallenbach NR (1983) Structure of apamin in solution, 2 dimensional NMR study. Biochemistry 22:1901-1906

# The Voltage-Sensitive Sodium Channel from Mammalian Skeletal Muscle

R. L. Barchi, R. E. Furman, and J. C. Tanaka[1]

## Introduction

In mammalian skeletal muscle, action potentials are produced by sequential changes in the conductance of the surface membrane to sodium and potassium ions. These variable conductances are modulated by separate voltage-sensitive ionic channels which span the bilayer and provide gated aqueous pathways for ion movement through the membrane. The characteristics of the currents controlled by the voltage-sensitive sodium channel in skeletal muscle have been the subject of intensive study since the pioneering work of Hodgkin and Huxley in the 1950's (Hodgkin and Huxley 1952). Early work in muscle relied on the traditional approach of voltage clamp to describe the kinetics of these sodium currents in large populations of channels in a single fiber (Adrian et al. 1970). More recently patch clamp technology has been used to extend this analysis to the behavior of single sodium channels in their native environment (Sigworth and Neher 1980).

During the past few years, rapid progress has been made in elucidating the biochemistry of a number of membrane ion channels; the skeletal muscle sodium channel has been no exception in this regard. Building on earlier work with the binding of radiolabeled toxins to the sodium channel in situ, this channel protein has been solubilized and purified from rat and rabbit skeletal muscle as well as from rat brain and eel electroplax. Although the purification of this protein on the basis of its toxin binding capacity represents an important step forward, it is the functional reconstitution that provides the necessary proof of its integrity as a membrane ion channel.

This discussion will deal specifically with the purification, reconstitution, and functional assessment of the sodium channel from mammalian skeletal muscle. Since the biochemical aspects of its purification have been the subject of other reviews (Barchi 1984, Barchi et al. 1984), they will be considered only briefly here. Sodium channels from brain and other sources are dealt with also in other reviews (Barchi 1984, Catterall 1984).

## Biochemical Characterization of Skeletal Muscle Sodium Channels

*Purification and Physical Properties.* The voltage-dependent sodium channel has been isolated from rat (Barchi et al. 1980, Barchi 1983) and rabbit (Kraner et al. 1985) skeletal muscle, and partially purified from human

[1]David Mahoney Institute of Neurological Sciences, the Departments of Neurology and of Biochemistry and Biophysics, University of Pennsylvania School of Medicine, Philadelphia, PA 19104, USA

36. Colloquium - Mosbach 1985
Neurobiochemistry

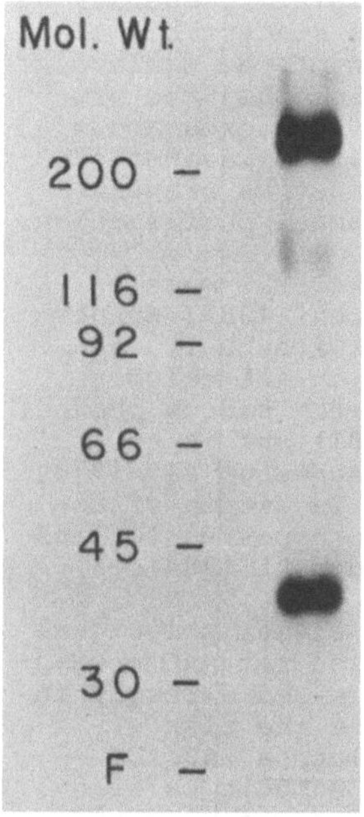


Fig. 1. Subunit composition of the purified sodium channel from rabbit skeletal muscle. The purified channel contains a large glycoprotein with an apparent MW of 260,000 and a smaller component of 38,000 MW. Proteins were labeled with Boltun-Hunter reagent prior to electrophoresis, separated on a 7 - 20% gradient SDS-polyacrylamide gel, and visualized by autoradiography

skeletal muscle (Roberts and Barchi 1985, unpublished results). The membrane isolation procedure used with rat muscle (Barchi et al. 1979) produces mostly sarcolemmal membranes while that used with rabbit muscle (Rosemblatt et al. 1981) generates predominantly T-tubular membranes. Comparable channel proteins have also been isolated from eel electric organ (Agnew et al. 1978, Miller et al. 1983, Norman et al. 1983), rat brain (Hartshorne and Catterall 1981, Hartshorne et al. 1982, Barhanin et al. 1983) and rat heart (Lombet and Lazdunski 1984).

The physical properties of the solubilized skeletal muscle sodium channel can be estimated by a combination of gel filtration and density gradient centrifugation (Barchi and Murphy 1980). When examined in this way, the rat, rabbit, and human skeletal muscle channels appear remarkably similar. Each exhibits an apparent Stokes radius of about 8.6 nm in its mixed micelle with detergent and phospholipid. The partial specific volume of the mixed micelle for each is about 0.83 ml/g. Using an estimate for the actual partial specific volume of the protein of 0.73 ml/g, the molecular weight of the channel protein in the absence of detergent and lipid was estimated to be about 300,000.

The rat sarcolemmal and rabbit T-tubular channels have been purified by chromatography on a weak anion exchange resin followed by affinity chromatography on a column of immobilized wheat germ agglutinin and centrifugation on a sucrose density gradient (Barchi et al. 1980, Barchi 1983, Kraner et al. 1985). Peak fractions from these sucrose gradients have exhibited saxitoxin binding activity in excess of 3000 pmol/mg protein. Assuming an average molecular weight for the total protein from physical measurements of 300,000, the theoretical maximal specific activity of the purified protein would be $\sim$3300 pmol/mg, suggesting that these final preparations easily exceed 90% in purity.

*Subunit Composition.* All three of the skeletal muscle sodium channels characterized in our laboratory contain a large glycoprotein subunit of ∿260 kDa that runs diffusely and with anomalous migratory characteristics on SDS-PAGE (Fig. 1). This subunit is quite susceptible to proteolytic nicking during the initial homogenization of the skeletal muscle and the subsequent preparation of membrane fractions; the products of this proteolytic activity are a diffuse band centered about 150 kDa and one smaller component of 45 kDa (Barchi 1983, Casadei et al. 1984, Casadei et al. 1985). Protease inhibitors must be present throughout the membrane isolations as well as the channel purification.

The purest preparations of each of these mammalian skeletal muscle channels also contain a small subunit of ∿38 kDa (Barchi 1983, Kraner et al. 1985). This subunit co-migrates with the 260 kDa subunit, as well as with the peak of functional channel activity on all columns and gradients. A similar 38 kDa component is seen in the sodium channel purified from rat brain (Hartshorne and Catterall 1981) but is not found in the channel isolated from eel electroplax (Agnew et al. 1978, Miller et al. 1983). Other workers have reported the isolation of sodium channels from rat brain and rat heart which also appeared to lack this component (Barhanin et al. 1983, Lombet and Lazdunski 1984).

We have used immunoaffinity columns constructed with either polyclonal or monoclonal antibodies generated against the purified rat sodium channel to isolate the channel protein from muscle homogenized directly in detergent and protease inhibitors in order to minimize the risk of proteolysis (Casadei et al. 1985). These columns isolate a 260 kDa protein from fresh rat muscle. In addition, a 38 kDa subunit is also clearly seen in spite of the fact that these antibodies are directed only against the 260 kDa subunit. The 38 kDA subunit coelutes from the affinity column with the 26 kDa component when thiocyanate gradients are used to displace bound protein from the covalently immobilized immunoglobulin. Although the 38 kDa subunit is sometimes difficult to detect on SDS-PAGE using silver stains due to its much smaller mass compared to the 260 kDa subunit, it is easily seen in these preparations when the proteins are iodinated under denaturing conditions prior to electrophoresis.

The role of the 38 kDa component remains to be resolved. In the case of the eel channel, its presence does not seem necessary for functional reconstitution of the channel (Rosenberg et al. 1984); in mammals the major polypeptide toxin binding site and the saxitoxin binding site appear to be located on the 260 kDa component (Beneski and Catterall 1980, Lombet et al. 1983). The 38 kDa component may represent an integral part of the mammalian channel; alternatively the peptide could be an associated regulatory protein, a cytoskeletal element linked to the channel, or an unrelated but persistant contaminant. In our experience, however, all rat or rabbit muscle channel preparations that exhibited functional activity after reconstitution have contained the 38 kDa component (Tanaka et al. 1983, Kraner et al. 1985).

## Reconstitution of the Purified Channel

*Flux Measurements with Reconstituted Sodium Channels.* The voltage-sensitive sodium channel has a number of biophysical and biochemical properties that uniquely define it among membrane ion channels. Saxitoxin binding is one such characteristic, and that property is used to identify the channel during its biochemical isolation; the binding does not, however, provide evidence of channel function. From electrophysiological and pharmacological studies a list of the properties that the reconstituted protein should possess can be derived. This list includes:

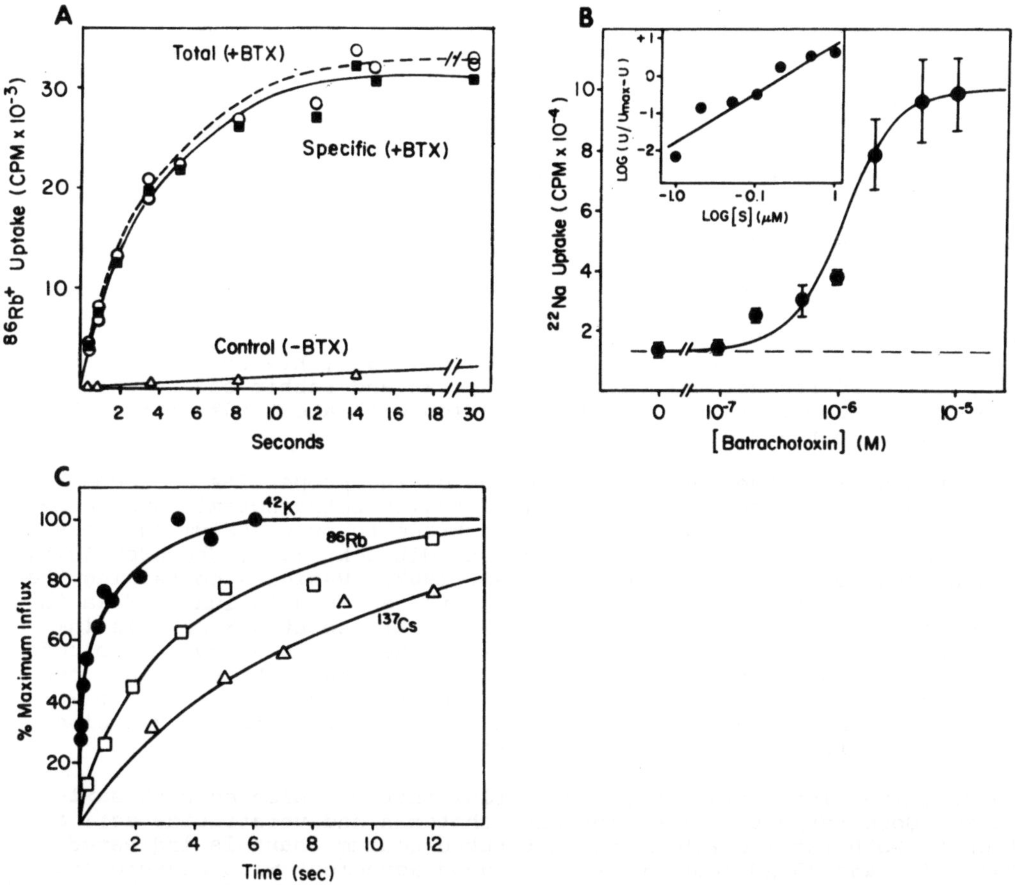


Fig. 2. (A) $^{86}Rb^+$ influx into vesicles containing the purified sodium channel. The purified channel was reconstituted into egg phosphatidylcholine vesicles and incubated for 45 min at 36°C with $5 \times 10^{-6}$ M batrachotoxin *(BTX)* (o) in ethanol (0.5% final concentration in the vesicle suspension) or with ethanol alone (Δ). Specific batrachotoxin-activated influx (■), defined as total minus control influx, occurred with a halftime of about 3 s in this example. Fluxes were measured after equilibration of the vesicles at 22°C. (B) Dose-dependent activation of the purified, reconstituted channel by batrachotoxin. Vesicles containing the purified channel were incubated for 45 min at 36°C with concentrations of batrachotoxin between $1 \times 10^{-7}$ and $1 \times 10^{-5}$ M prepared in ethanol or with an equivalent amount of ethanol alone. Influx of $^{22}Na^+$ was then measured after 15 s, a period sufficient for all vesicles containing opened channels to equilibrate with this cation. *Inset* Hill plot of the same data indicating an apparent $K_d$ of 2 μM and a Hill coefficient of 1.2. (C) Relative rate of uptake of monovalent cations into vesicles containing the reconstituted channel activated by batrachotoxin. The total volume accessible to each cation through activated channels was comparable, and uptake is expressed as percent maximal uptake after correction for nonspecific influx. Measurements were made at 22°C following activation of vesicles with batrachotoxin ($5 \times 10^{-6}$ M at 36°C for 45 min). ●, $^{42}K^+$; □, $^{86}Rb^+$; Δ, $^{137}Cs^+$. $^{22}Na^+$ uptake was complete prior to 100 ms, the earliest time point resolvable (adapted from Kraner et al. 1985)

(1) the appropriate toxin pharmacology, (2) the ability to select for sodium among similar monovalent cations, (3) a characteristic maximal rate of ion flow or single channel conductance, and (4) characteristic voltage-dependent channel transitions between conducting and nonconducting states. Proof that the purified saxitoxin binding protein is the voltage-dependent sodium channel demands evidence for as many of these channel properties as possible with the purified protein after reconstitution.

Sodium channel protein purified from rat and rabbit skeletal muscle has been reconstituted into egg phosphatidylcholine vesicles using BioBeads SM2 to remove the NP-40 detergent. In order to assess channel orientation, saxitoxin binding was measured on reconstituted fractions before and after the addition of a small amount of detergent to disrupt membrane integrity (Weigele and Barchi 1982, Kraner et al. 1985). In routine PC reconstitutions, both the rat and rabbit sodium channels inserted randomly during vesicle formation with about 50% of the channels facing outward.

Neurotoxins were then used to probe the gating properties of reconstituted sodium channels. The alkaloid toxins batrachotoxin and veratridine act from either side of the membrane to open sodium channels and prevent or delay their inactivation (Ulbricht 1969, Khodorov 1978, Quandt and Narahashi 1982, Huang et al. 1982). Vesicles containing the purified reconstituted channel demonstrated a rapid influx of $^{22}$Na in response to batrachotoxin activation (Fig. 2). About 50% of this influx was inhibited by external saxitoxin. The remaining 50% was blocked by including saxitoxin in the medium trapped within the vesicles, consistent with the known random orientation of the channel in the bilayer. Thus all channels capable of activation by batrachotoxin also contained a functional binding site for saxitoxin.

Veratridine also activated a $^{22}$Na influx that was blocked with saxitoxin. Dose response curves for batrachotoxin and veratridine activation of both rat and rabbit reconstituted sodium channels indicated $K_d$'s of 5 and 35 µM respectively, in good agreement with measurements on native channels. The $K_i$ for saxitoxin block (measured with $^{86}$Rb uptake at 5 s) was 5 nM in batrachotoxin-activated vesicles. Taken as a whole, these results demonstrate a pharmacological profile for the reconstituted saxitoxin binding protein that is comparable to the native sodium channel (Catterall 1977) and confirm that the reconstituted proteins modulate $^{22}$Na membrane permeability in response to alkaloid activation.

*Cation Selectivity.* The timecourse for the influx of various cations into batrachotoxin-activated vesicles containing the purified channel was measured with a quenched flow apparatus that allowed clear resolution of $^{42}$K, $^{86}$Rb, and $^{137}$Cs equilibration rates (Tanaka et al. 1983, Kraner et al. 1985). The early linear phase of $^{42}$K uptake was used to calibrate the resolution of the system; extrapolation to zero uptake indicated a 90 ms deadtime during which uptake could not be measured accurately. For both the rat and rabbit channels, $^{22}$Na influx was complete prior to 90 ms, but an upper limit of 50 ms could be placed on the halftime for $^{22}$Na equilibration. Using this estimate, selectivity ratios were calculated for the other cations. For the purified rabbit channel these values were $Na^+$ (1.0) : $K^+$ (0.13) : $Rb^+$ (0.02) : $Cs^+$ (0.008) (Kraner et al. 1985). Virtually identical results were obtained with the purified channel from rat sarcolemma (Tanaka et al. 1983). Although there are reports that batrachotoxin reduces the selectivity of the sodium channel (Khodorov 1978, Huang et al. 1979, Frelin et al. 1981), the reconstituted channel clearly retains much of its native selectivity (Hille 1972, Pappone 1980).

Cation uptake into veratridine-activated vesicles occurred several orders of magnitude more slowly than that seen in the same vesicles after batrachotoxin activation. The selectivity of the veratridine-modified channel was the same as that of native or batrachotoxin-modified channels, but relative selectivity among cations appeared greatly reduced.

The slow equilibration rates seen with veratridine may reflect the relatively infrequent channel openings produced by this toxin as compared to batrachotoxin, while part of the apparent loss in selectivity may be due to the longer duration of these opening events when they do occur. Thus for veratridine activated channels, channel opening rather than ion movement through the open channel may be the rate-limiting step.

## Voltage-Dependent Toxin Activation

*Single Channel Measurements.* We initially observed voltage-dependent and pharmacological behavior consistent with native sodium channels with inside-out patches from reconstituted vesicles containing rat skeletal muscle sodium channels purified in CHAPS (Barchi et al. 1984). Single channel openings were recorded as a function of transmembrane potential in a batrachotoxin activated inside-out patch. Hyperpolarization shifted the channel from a mostly open to mostly closed state. The percent open time versus membrane potential curve was shifted in the hyperpolarizing direction as observed with batrachotoxin-activated sodium channels in situ (Huang et al. 1982). Some patches showed spontaneous, infrequent burst openings with depolarization prior to the addition of batrachotoxin. Immediately after batrachotoxin activation, the opening bursts converted to continuous channel openings and closings.

More recently, we have studied the sodium channel from rabbit T-tubular membranes purified in NP-40 by fusing vesicles containing these channels with planar bilayers (Furman et al. 1985). Bilayers were formed across a 350 μM orifice using 1-palmitoyl, 2-oleoyl PE and 1-palmitoyl, 2-oleoyl PC (80:20) in decane (50 mg lipid/ml). The solution on the cis side of the membrane contained 500 mM NaCl, 10 mM HEPES, pH 7.4, 0.15 mM $CaCl_2$, 0.1 mM $MgCl_2$, and 0.05 mM EGTA while the trans side substituted 200 mM NaCl in the solution.

Sodium channels generally incorporated into the bilayer with their extracellular tetrodotoxin site facing the trans chamber. The single channel conductance averaged ∿20 pS and the current-voltage relation was linear over the range examined. The channel activation was steeply voltage-dependent with the 50% opening point at -95 mV (Furman et al. 1985). A shift of 5 mV about the midpoint dramatically altered the ratio of times spent in the open and closed states (Fig. 3). Application of tetrodotoxin to the cis chamber (intracellular side of the channel) had no effect on channel activity, but applying the toxin to the trans (extracellular) side closed the channel for greater than 99% of the time. Similar results have been reported for the purified rat brain sodium channel reconstituted into planar bilayers (Hartshorne et al. 1985).

*Ensemble Measurements.* Single channel analysis of purified sodium channels leaves unresolved the question of whether those few channels observed in the bilayer are truly representative of all the channel protein purified. The single channel conductance, voltage dependence and sensitivity to tetrodotoxin seen with these purified single channels

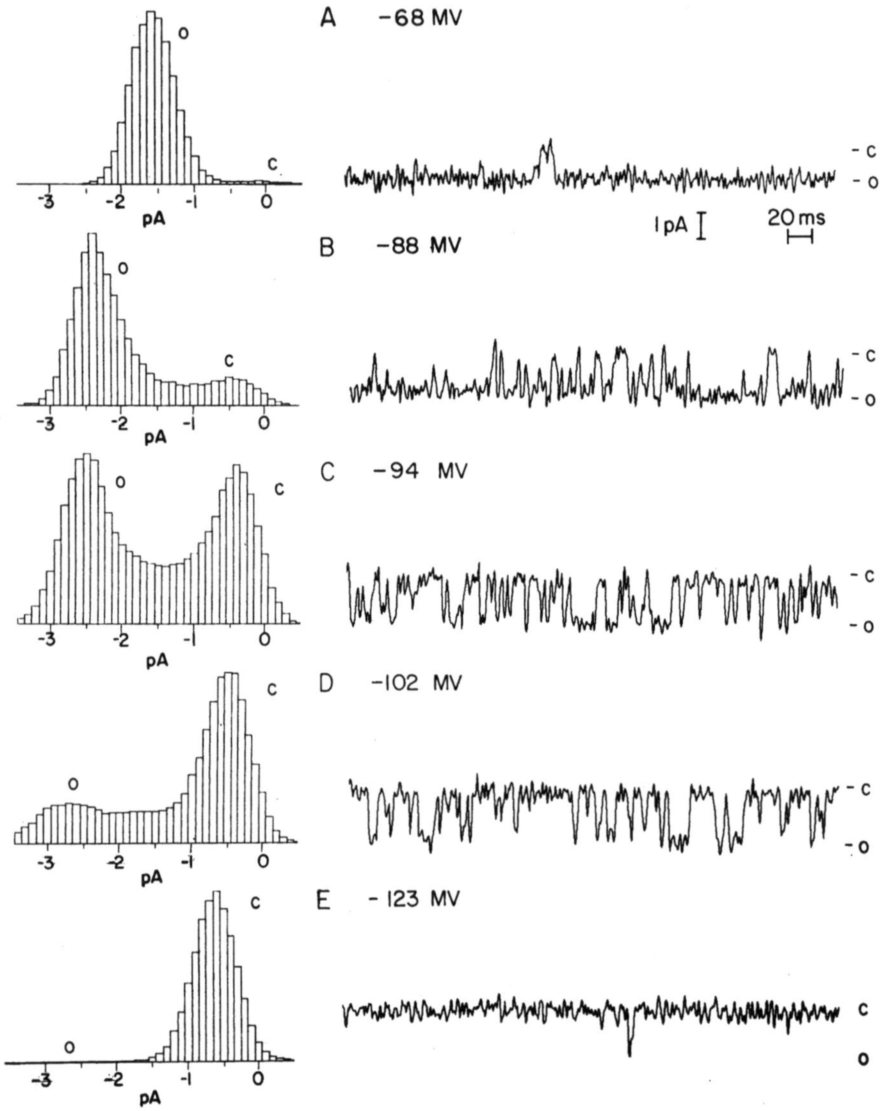


Fig. 3. Amplitude histograms and corresponding current records from a purified rabbit sodium channel reconstituted into a planar bilayer illustrating the voltage dependence of channel gating. Hyperpolarization (A to E) increasingly shifts the channel from the open (O) to closed states (C). The point where the batrachotoxin activated channel was open 50% of the time was about -95 mV (physiological convention; extracellular side of channel was trans). Single channel conductance was 22 pS. Current records are filtered at 400 Hz. Bilayers were formed as described in the text. Cis solution: 500 mM NaCl, 10 mM HEPES, 0.15 mM $CaCl_2$, 0.1 mM $MgCl_2$, 0.05 mM EGTA, pH 7.4. Trans solution was the same except for 200 mM NaCl and $1 \times 10^{-6}$ M batrachotoxin. Temperature was 22°C. (Adapted from Furman et al. 1986)

in planar bilayers closely resemble those previously reported for native sodium channels. However, channels that do not fuse with the bilayer or are not activated contribute no information in these studies. This issue can be addressed by the measurement of voltage-dependent activation of large populations of purified channels in the reconstituted vesicle system.

Depolarization of a sodium channel in situ normally leads to a brief increase in the membrane sodium conductance that lasts about 1 ms before the channel inactivates (Sigworth and Neher 1980, Horn et al. 1981). If reconstituted sodium channels could be activated in vesicles, this 1 - 2 ms open time would be too brief to allow sufficient isotope influx to be measured accurately. As confirmed in the preceding single-channel studies, sodium channels activated by batrachotoxin or veratridine retain their voltage-dependent activation although the relationship between activation and voltage is shifted about 40 mV in the hyperpolarizing direction (Huang et al. 1982, Leibowitz et al. 1985). The other major action of these toxins is to slow or eliminate the inactivation process so that channels remain opened with prolonged depolarization. Voltage-dependent channel gating should be detectable in vesicles containing batrachotoxin-activated channels if the vesicle membrane potential can be controlled.

In order to establish a membrane potential in the reconstituted vesicles, a potassium concentration gradient was created across the vesicle membrane and the vesicles were rendered permselective to $K^+$ by the addition of valinomycin. Vesicles were formed in KCl buffer with 2 μM saxitoxin trapped internally to block inward-facing channels and create a functionally uniform population of outward-oriented channels. External saxitoxin and KCl were then removed and replaced with choline chloride. Following the addition of $10^{-7}$ M valinomycin (Andreoli et al. 1967, Mueller and Rudin 1967), the hyperpolarized vesicles were activated with batrachotoxin.

The voltage dependence of these batrachotoxin-activated channels was determined by measuring $^{22}Na$ uptake at various membrane potentials set by altering external $|K|^+$. We found that the specific $^{22}Na$ flux was maximal at low membrane potentials ($K_o = K_i$) and completely inhibited a large hyperpolarizing potentials ($K_o = 0$) (Furman et al. 1985). The flux increased sharply over a range of external potassium concentrations corresponding to predicted Nernst potentials between -100 and -60 mV, the general range observed for batrachotoxin modified channels in situ.

These experiments clearly demonstrate that all the batrachotoxin-activated and saxitoxin inhibitable sodium channels present in the reconstituted vesicle population retain at least one of the voltage-sensitive properties characteristic of this channel and that the general relationship between membrane potential and channel activation was comparable to batrachotoxin-treated channels in their native membrane. Although this approach lacks the temporal resolution of single-channel recording, and cannot provide information on channel kinetics, it yields insight into the response of the entire channel population that is not available from patch clamping and planar bilayer measurements.

## Summary

On the basis of reconstitution studies that allow direct measurement of cation fluxes and single channel properties, we can state with certainty that the protein originally purified from skeletal muscle as a saxitoxin binding component is in fact the voltage-sensitive sodium ion channel. In its purified form this channel retains its capacity to gate cation currents in response to pharmacologic activation and block, to select sodium ions among other cations of similar charge, and to respond dramatically to changes in membrane potential. Its single channel properties are comparable to those of a native channel, and flux measurements confirm that the voltage-dependent gating seen with single channels extends to all the channels that are capable of batrachotoxin activation and saxitoxin block.

Reconstitution has so far provided no insight into the role of the 38 kDa component of the mammalian sodium channel since this band was present in all our rat and rabbit purified preparations. Our reconstitution studies have also been unable to provide evidence for or against skeletal muscle sodium channel subtypes. These and other issues will certainly be addressed in the near future. The recent elucidation of the primary sequence of the eel sodium channel (Noda et al. 1984) and the active work now underway to clone both the mammalian muscle and brain channels point the way toward the next major chapter in the molecular study of this ion channel, the detailed correlation of structure with function.

*Acknowledgments.* Work described in this chapter was supported in part by NIH grants NS-08075 and NS-18013 and by a grant from the Muscular Dystrophy Association.

## References

Adrian RH, Chandler WK, Hodgkin AL (1970) Voltage clamp experiments in striated muscle fibers. J Physiol 208:607-644

Agnew WS, Levinson SR, Brabson JS, Raftery MA (1978) Purification of the terodotoxin-binding component associated with the voltage-sensitive sodium channel from *Electrophorus electricus* electroplax membranes. Proc Natl Acad Sci USA 75:2606-2610

Andreoli TE, Tieffenberg M, Tosteson DC (1967) The effect of valinomycin on the ionic permeability of thin lipid membranes. J Gen Physiol 50:2527-2545

Barchi RL (1983) Protein components of the purified sodium channel from rat skeletal muscle sarcolemma. J Neurochem 40:1377-1385

Barchi RL (1984) Voltage-sensitive $Na^+$ ion channels: molecular properties and functional reconstitution. TIBS 9:358-361

Barchi RL, Murphy LE (1980) Size characteristics of the solubilized sodium channel saxitoxin binding site from mammalian sarcolemma. Biochim Biophys Acta 597:391-398

Barchi RL, Weigele J, Chalikian D, Murphy L (1979) Muscle surface membranes. Preparative methods affect apparent chemical properties and neurotoxin binding. Biochim Biophys Acta 550:59-76

Barchi RL, Cohen SA, Murphy LE (1980) Purification from rat sarcolemma of the saxitoxin-binding component of the excitable membrane sodium channel. Proc Natl Acad Sci USA 77:1306-1310

Barchi RL, Tanaka JC, Furman RF (1984) Molecular characteristics and functional reconstitution of muscle voltage-sensitive sodium channels. J Cell Biochem 26: 135-146

Barhanin J, Pauron D, Lombet A, Norman RI, Vijverberg PM, Giglio JR, Lazdunski M (1983) Electrophysiological characterization, solubilization and purification of the *Tityus* toxin receptor associated with the gating component of the $Na^+$ channel from rat brain. EMBO J 2:915-920

Beneski DA, Catterall WA (1980) Covalent labeling of protein components of the sodium channel with a photoactivable derivative of scorpion toxin. Proc Natl Acad Sci USA 77:639-643

Casadei JM, Gordon RD, Lampson LA, Shotland DL, Barchi RL (1984) Monoclonal antibodies against the voltage-sensitive $Na^+$ channel from mammalian skeletal muscle. Proc Natl Acad Sci USA 81:6227-6231

Casadei JM, Gordon RD, Barchi RL (1986) Immunoaffinity purification of the voltage dependent sodium channel from mammalian skeletal muscle. J Biol Chem, in press

Catterall WA (1977) Activation of the action potential $Na^+$ ionophore by neurotoxins. J Biol Chem 252:8669-8676

Catterall WA (1984) The molecular basis of neuronal excitability. Science 223:653-661

Frelin C, Vigne P, Lazdunski M (1981) The specificity of the sodium channel for monovalent cations. Eur J Biochem 119:437-442

Furman RE, Tanaka J, Barchi RL (1986) Voltage-dependent activation of purified sodium channels from rabbit T-tubular membranes. Proc Natl Acad Sci USA 83:488-492

Hartshorne RP, Catterall WA (1981) Purification of the saxitoxin receptor of the sodium channel from rat brain. Proc Natl Acad Sci USA 78:4620-4624

Hartshorne RP, Messner DJ, Coppersmith JC, Catterall WA (1982) The saxitoxin receptor of the sodium channel from rat brain. Evidence for two nonidentical B subunits. J Biol Chem 257:13888-13891

Hartshorne R, Keeler BU, Talvenheimo JA, Catterall WA, Montal M (1985) Functional reconstitution of the purified brain sodium channel in planar lipid bilayers. Proc Natl Acad Sci USA 82:240-244

Hille B (1972) The permeability of the sodium channel to metal cations in myelinated nerve. J Gen Physiol 59:637-658

Hodgkin AL, Huxley AF (1952) The components of membrane conductance in the giant axon of Loligo. J Physiol (London) 116:473-496

Horn R, Patlak J, Stevens CF (1981) Sodium channels need not open before they inactivate. Nature (London) 291:426-427

Huang LM, Catterall WA, Ehrenstein G (1979) Comparison of ionic selectivity of batrachotoxin-activated channels with different tetrodotoxin dissociation constants. J Gen Physiol 73:839-854

Huang LM, Moran N, Ehrenstein G (1982) Batrachotoxin modifies the gating kinetics of sodium channel in internally perfused neuroblastoma cells. Proc Natl Acad Sci USA 79:2082-2085

Khodorov BI (1978) Chemicals as tools to study nerve fiber sodium channels; effects of batrachotoxin and some local anesthetics. In: Tosteson DC, Ovchennikov AY Latorre R (eds) Membrane Transport Process, vol II. Raven Press, New York

Kraner SD, Tanaka JC, Barchi RL (1985) Purification and functional reconstitution of the voltage-sensitive sodium channel from rabbit T-tubular membranes. J Biol Chem 260:6341-6347

Leibowitz MD, Sutro JB, Hille B (1985) Four lipid-soluble toxins modify sodium channel gating. Biophys J 47:32a

Lombet A, Lazdunski M (1984) Characterization, solubilization, affinity labelling and purification of the cardiac sodium channel using *Tityus* toxin. Eur J Biochem 141:651-660

Lombet A, Norman R, Lazdunski M (1983) Affinity labelling of the tetrodotoxin binding component of the $Na^+$ channel. Biochem Biophys Res Commun 114:126-130

Miller JA, Agnew WS, Levinson SR (1983) Principal glycopeptide of the tetrodotoxin saxitoxin binding protein from *Electrophorus electricus*; Isolation and partial chemical and physical characterization. Biochemistry 22:462-470

Mueller P, Rudin DO (1967) Development of $K^+$-$Na^+$ discrimination in experimental bimolecular lipid membranes by macrocyclic antibodies. Biochem Biophys Res Commun 26:398-404

Noda M, Shimizu S, Tsutomu T, Takai T, Kayano T, Ikeda T, Takahashi H, Nakayama H, Kanaoka U, Minamino N, Kangawa K, Matsuo H, Raftery MA, Hirose T, Inayama S, Hayashida H, Miyata T, Numa S (1984) Primary structure of *Electrophorus electricus* sodium channel deduced from cDNA sequence. Nature (London) 312:121-127

Norman RI, Schmidt A, Lombet A, Barhanin J, Lazdunski M (1983) Purification of binding protein from *Tityus* toxin identified with the gating component of the voltage-sensitive $Na^+$ channel. Proc Natl Acad Sci USA 80:4164-4168

Pappone P (1980) Voltage-clamp experiments in normal and denervated mammalian skeletal muscle fibers. J Physiol (London) 306:377-410

Quandt FN, Narahashi T (1982) Modification of single $Na^+$ channels by batrachotoxin. Proc Natl Acad Sci USA 79:6732-6736

Redfern P, Thesleff S (1971) Action Potential generation in denervated rat skeletal muscle. Acta Physiol Scand 81:557-564

Rosemblatt M, Hidalgo C, Vergara C, Ikemoto N (1981) Immunological and biochemical properties of transverse tubule membranes isolated from rabbit skeletal muscle. J Biol Chem 256:8140-8148

Rosenberg RL, Tomiko SA, Agnew WS (1984) Reconstitution of neurotoxin-modulated ion transport by the voltage-regulated sodium channel isolated from the electroplax of *Electrophorus electricus*. Proc Natl Acad Sci USA 81:1239-1243

Sherman SJ, Lawrence JC, Messner DJ, Jacoby K, Catterall WA (1981) Tetrodotoxin-sensitive sodium channels in rat muscle cells developing in vitro. J Biol Chem 258:2488-2495

Sigworth FJ, Neher E (1980) Single $Na^+$ channel currents observed in cultured rat muscle cells. Nature (London) 287:447-449

Tanaka JC, Eccleston JF, Barchi RL (1983) Cation selectivity characteristics of the reconstituted voltage-dependent sodium channel purified from rat skeletal muscle sarcolemma. J Biol Chem 258:7519-7526

Ulbrich W (1969) The effect of veratridine on excitable membranes of nerve and muscle. Ergeb Physiol Biol Chem Exp Pharmacol 61:18-71

Weigele JB, Barchi RL (1982) Functional reconstitution of the purified sodium channel protein from rat sarcolemma. Proc Natl Acad Sci USA 79:3651-3655

# Biochemistry and Physiology of Cardiac Calcium Channels

F. Hofmann, V. Flockerzi, J. Oeken, and P. Ruth[1]

## Introduction

The voltage-dependent slow calcium channel allows the passive movement of calcium across the plasma membrane of a variety of excitable cells and regulates thereby directly or indirectly the cytosolic calcium concentration of these cells (for a review see Reuter 1984). In cardiac muscle, opening of these channels depends on the depolarization of the membrane and is facilitated by phosphorylation of a membrane protein (Osterrieder et al. 1982). In vascular smooth muscle and in other tissues, a second type of calcium channel has been postulated which is regulated directly or indirectly by hormone receptors. These channels appear to differ in many respects from the voltage-operated calcium channels. Recently, a third type of calcium channel has been described which differs markedly in its electrophysiological characteristics from the other two types (Carbone and Lux 1984, Armstrong and Matteson 1985). The calcium conductance of the voltage- and receptor-operated channel is blocked by a heterogenous group of compounds, for which the term "calcium antagonists" (Fleckenstein 1977), calcium entry blockers, or calcium channel blockers has been introduced. Among these compounds are the phenylalkylamines verapamil, gallopamil, and desmethoxyverapamil, the dihydropyridines nitrendipine, nimodipine, and nifedipine, and the benzothiazepine diltiazem. These compounds are used therapeutically in various cardiovascular diseases. Their tritiated congeners have been employed as valuable tools to identify calcium channels in broken cell preparations (Glossmann et al. 1982). This article will summarize some aspects of the cardiac voltage-operated calcium channel.

## In Vivo Modification of the Calcium Channel

During the last decade, a large body of evidence has been accumulated which suggested that stimulation of the β-adrenergic receptor increases the amount of calcium entering the cardiac muscle cell during depolarization. It was proposed that this change in calcium conductance is caused by cAMP-dependent phosphorylation of the calcium channel or a protein closely related to the channel. This prediction was later confirmed by experiments in which the catalytic subunit of cAMP-dependent protein kinase was injected into single cardiac muscle cells (Osterrieder et al. 1982). Injection of the catalytic subunit of cAMP-dependent protein kinase increased by about threefold the calcium conductance of voltage-clamped cells. These experiments ruled out the possibility that cAMP may modify the calcium conductance by an indirect mechanism, such as changing the potassium permeability. However, in

[1]Physiologische Chemie, Medizinische Fakultät der Universität des Saarlandes, D-6650 Homburg/Saar, FRG

36. Colloquium Mosbach 1985
Neurobiochemistry

these initial experiments rather high concentrations of the protein kinase were used. This raised the possibility that the observed increase in the calcium conductance is caused by nonphysiological concentrations of the enzyme and is not relevant to the physiological situation. Recent experiments (Kameyama et al. 1985) in which single cardiac cells were perfused with various concentrations of the catalytic subunit of cAMP-dependent protein kinase show that already 1 μM subunit increases half-maximally the calcium conductance (Table 1). This enzyme concentration is almost identical to the concentration of catalytic subunit measured in cardiac muscle (Hofmann et al. 1977). The calcium conductance was raised maximally 3.3-fold by 10 μM catalytic subunit. Similar changes are obtained either by perfusion of a single cell with cAMP or by the addition of a maximal concentration of isoproterenol to the perfusion chamber, suggesting that the catalytic subunit elicited the physiological response at a physiological concentration. The phosphorylation of the cardiac calcium channel increases the opening probability of the channel during depolarization (Hess et al. 1984) by decreasing the number of zero sweeps during single channel recording. In an intact cell, the change in the opening probability increases the number of channels which are open during depolarization and therefore enhances the calcium conductance. Perfusion of a single cell with cAMP-free regulatory subunit of cAMP-dependent protein kinase decreases the calcium conductance, if the cell is stimulated by isoproterenol (Osterrieder et al. 1982). This indicates that removal of free catalytic subunit increases the rate of dephosphorylation of the channel and thereby decreases the probability of its opening. However, the calcium conductance cannot be abolished completely by perfusion of a cell with high concentration of regulatory subunit. Roughly, about one third of the total number of calcium channels can still be opened in the absence of cAMP-dependent phosphorylation, indicating that two functionally different populations of channels are present in cardiac cell, namely phosphorylated and nonphosphorylated ones.

Table 1. Effect of isoproterenol, cAMP, and catalytic subunit of cAMP-kinase (C-subunit) on slow calcium current in single, perfused cardiac myocyte

| Compound | $K_A$ | $I_{Ca}$ |
|---|---|---|
| | μM | Stimulation (fold) |
| None | - | 1 |
| Isoproterenol | 0.04 | 3.7 |
| cAMP | 5 | 3.5 |
| C-Subunit | 1 | 3.3 |

Compounds were applied either intracellularly continuous perfusion (cAMP; C-subunit) or were added directly to the perfusion chamber. The membrane current was measured by using single electrode voltage clamp technique. For further details see Kameyama et al. (1985). $K_A$, concentration to stimulate half-maximally the slow inward calcium current ($I_{Ca}$). The concentration refers to that present in the perfusion pipette or added to the bath.

## Measurement of Calcium Fluxes in Cardiac Sarcolemmal Vesicles

The experiments carried out in vivo suggested that phosphorylation of a membrane protein changed the properties of the calcium channel. To identify the protein responsible for these changes in vitro studies are required which mimic the in vivo situation. According to Bartschat et al. (1980), potassium/sodium polarized vesicles of cardiac sarcolemma accumulate "voltage"-dependent calcium if the extravesicular potassium concentration is raised (see Fig. 1). Using this technique it was reported that phosphorylation of a 22 kDa peptide named calciductin by cAMP-dependent protein kinase increases in vitro the "voltage"-dependent calcium uptake of the sarcolemmal vesicles (Rinaldi et al. 1982). In contrast to this report, a number of protein kinases phosphorylate in highly purified sarcolemma a 22 kDa peptide which is indistinguishable from phospholamban, a peptide of the sarcoplasmic reticulum (Fig. 1). Phosphorylation of this peptide does not increase

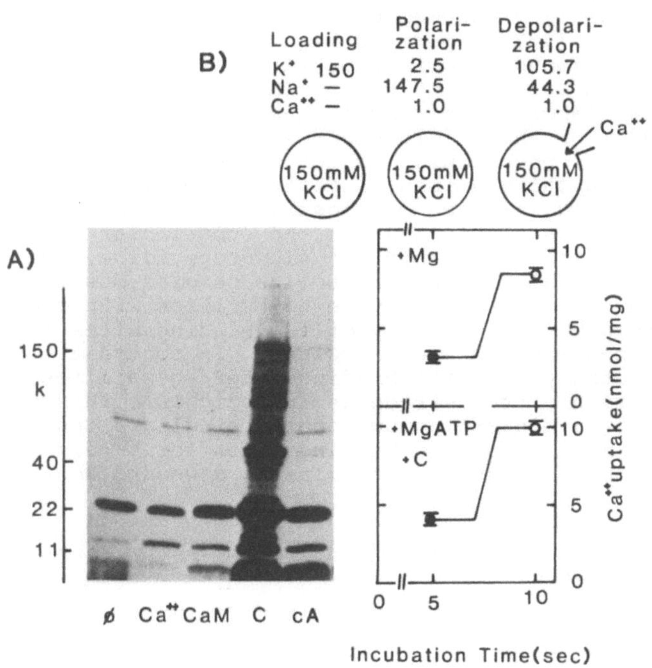


Fig. 1 A,B. Phosphorylation of cardiac sarcolemma and potassium, "depolarization"-induced calcium uptake. A Bovine cardiac sarcolemma (50 µg protein) was phosphorylated at 37°C for 2 min in the presence of [$\gamma$-$^{32}$P]•ATP Mg and following further additions: none, ø; 0.2 mM calcium (*$Ca^{2+}$*); 0.2 mM calcium and 5.8 µM calmodulin (*CaM*); 0.6 µM catalytic subunit of cAMP-kinase (*C*); or 10 µM cAMP (*cA*). The autoradiogram was developed after electrophoresis of the samples on a 10% to 22% SDS-gel. B Depolarization-induced calcium uptake was measured under the conditions shown in the upper part of the figure. The lower part shows the amount of $^{45}$Ca associated with vesicles after dilution into 0.1 mM lanthanum and washing them extensively on a membrane filter. Values are the mean ± SEM for 10-20 different determinations per point. Note that the difference in calcium uptake due to depolarization is identical for vesicles treated with 1 mM magnesium or 1 mM magnesium, 0.3 mM ATP, and 0.5 µM catalytic subunit (*C*). The total intravesicular volume was 20 µl/mg protein. For further details see Flockerzi et al. (1983)

the amount of calcium taken up by these vesicles after "depolarization" (Flockerzi et al. 1983). The difference between the results of both groups is more apparent than real, since in the experiments of Rinaldi et al. the catalytic subunit of cAMP-dependent protein kinase probably hydrolysed ATP to adenosine and phosphate. The phosphate precipitated together with calcium and lanthanum on the filter. The study by Flockerzi et al. avoided this artifact and showed that the amount of calcium taken up by the vesicles is within the range which can passively enter these vesicles. However, this calcium uptake was not inhibited by the organic calcium channel blockers, suggesting that the method of Bartschat et al. is not sufficient to identify in vitro cardiac calcium channels.

## Identification of Calcium Channels by Radioactive Ligands

Recently, the tritiated congeners of the organic calcium channel blockers have been used to identify calcium channels in broken cell preparations (Glossmann et al. 1982). Specific high affinity sites for these compounds have been found in different tissues including skeletal muscle, heart, smooth muscle, and brain. These binding sites have been studied in several laboratories and it has been suggested that two (Murphy et al. 1983) or three (Glossmann et al. 1985) separate sites exist in broken cell preparations which are specific for the dihydropyridines, the phenylalkylamines, and the benzothiazepine diltiazem. The physiological significance of these binding sites has not been established beyond doubt. Table 2 shows the subcellular distribution of the binding sites for nimodipine, a dihydropyridine, and (-)desmethoxyverapamil, a phenylalkylamine, in bovine cardiac muscle. The majority of the binding sites for nimodipine co-purifies with the sarcolemmal fraction, whereas only a minority of the binding sites for (-)desmethoxyverypamil is localized in the sarcolemma. In contrast, the majority of the (-)desmethoxyverapamil binding sites co-purifies with a fraction containing mainly free sarcoplasmic reticulum. This latter fraction contains only a minimal amount of the binding sites for nimodipine. Saturation analysis of the binding sites for (-)desmethoxyverapamil in both membrane fractions results in downward con-

Table 2. Distribution of binding sites for phenylalkylamines and dihydropyridines in different membranes as cardiac muscle

| | (-)D888 | Nimodipine | Strophantin | $Ca^{2+}$-Uptake |
|---|---|---|---|---|
| | fold (pmol/mg prot) | | (μmol/min × mg) | |
| Homogenate | 1 (.04) | 1 (.035) | 1 (3.3) | 1 (.01) |
| Sarcolemma | 3.4 | 10 | 10.7 | n.d. |
| Heavy SR | 11 | 3.2 | 0.9 | 48 |

Membranes were prepared from fresh bovine cardiac muscle as described by Flockerzi et al. (1983) and Jones and Cala (1981). The concentration of radioligand was 1 nM [Nimodipine, (-)desmethoxyverapamil (-D888)] and 1 μM (strophantin). For further details see Ruth et al. (1985). Values are the average of 4 to 20 different membrane preparations.

cave Scatchard plots (Ruth et al. 1985), suggesting the presence of two separate binding sites. Calculation of the data by a computer program for a two-site Scatchard plot indicates that each membrane fraction contains a high and a low affinity site with $K_D$ values in the nmolar and µmolar range, respectively (Table 3). A similar analysis for the sarcolemmal binding sites for nimodipine again reveals the existence of a high and low affinity site (Table 3). Nimodipine binds optimally to the high affinity site only in the presence of mmolar calcium, whereas the binding of (-)desmethoxyverapamil to its high and low affinity sarcolemmal sites is inhibited by a free calcium concentration of above 1 mM (Table 4). The high affinity sites for both ligands are specific for the respective group of calcium channel

Table 3. Kinetic constants of the cardiac muscle binding sites for calcium channel blockers

| Compound | High affinity site | | Low affinity site | |
|---|---|---|---|---|
| | $K_D$ | $B_{max}$ | $K_D$ | $B_{max}$ |
| Free sarcoplasmic reticulum | | | | |
| (-)Desmethoxyverapamil | 8.0 | 0.45 | 210 | 26 |
| Nimodipine | n.m. | n.m. | n.m. | n.m. |
| Sarcolemma | | | | |
| (-)Desmethoxyverapamil | 1.4 | 0.16 | 171 | 13.6 |
| Nimodipine | 0.35 | 0.30 | 33 | 8.2 |

Values are the average of several determinations using two or more different membrane fractions. Values were calculated by Scatfit computer programm. $K_D$ and $B_{max}$ values are given in nM and pmol/mg protein, respectively. *n.m.* not measurable.

Table 4. Summary of the properties of the cardiac muscle binding sites for calcium channel blockers

| | Phenylalkylamine | Dihydropyridine |
|---|---|---|
| Apparent localization | Sarcopl. reticulum | - |
| | Sarcolemma | Sarcolemma |
| High and low affinity sites | Yes | Yes |
| Calcium requirement of | | |
| High affinity site | No | Yes |
| Low affinity site | No | No |
| Binding affected by | | |
| Phenylalkylamines | Competitive | Allosteric |
| Dihydropyridines | Allosteric | Competitive |
| Diltiazem | Noncompetitive | Allosteric |

blockers and select with some exceptions the (-)isomers over the (+)-isomers. However, as shown by Ruth and coworkers, binding of nimodipine is allosterically modified by occupancy of the phenylalkylamine site and also by the occupancy of a site specific for d-cis-diltiazem. Binding of a dihydropyridine to its high affinity site inhibits allosterically the binding of (-)desmethoxyverapamil. In addition, binding of (-)desmethoxyverapamil is inhibited by diltiazem by a noncompetitive mechanism, suggesting that binding to the high affinity site of each group of calcium channel blockers affects allosterically the binding of the other two groups. These results indicate that cardiac muscle contains at least a separate site for each group of calcium channel blockers.

## Identity of the Sarcolemmal Binding Sites for Calcium Channel Blockers

As expected, cardiac muscle contains a number of binding sites for calcium channel blockers, the identity of which is questionable. The phenylalkylamine binding sites localized in the free sarcoplasmic reticulum are probably not related to that calcium channel which opens during depolarization and through which calcium enters the cell. This latter channel is usually studied in electrophysiological experiments and was modified by cAMP-dependent phosphorylation. The sarcoplasmic reticulum binding sites for (-)desmethoxyverapamil may be identical with that channel which is responsible for the release of calcium from the sarcoplasmic reticulum during muscle contraction. The binding sites localized in the sarcolemma could represent part of the voltage-operated calcium channel. Recently, Millard et al. (1983) suggested that the low affinity binding site for nimodipine is localized on the cardiac calcium channel and that binding to this site blocks the channel in vivo. However, comparison of the density of this site (about 1 pmol/mg muscle homogenate protein) with that of the number and density of functional calcium channels present in a cardiac cell (Table 5) suggests that there would be roughly 100 times more binding sites than channels. Such a large discrepancy seems to be rather unlikely. In

Table 5. Possible function of cardiac muscle binding sites for calcium channel blockers

| | |
|---|---|
| Number of functional $Ca^{2+}$ channels | 3 - 16 $10^3$/cell[a] |
| Density of functional $Ca^{2+}$ channels | 13 fmol/mg protein[c] |
| Density of high affinity DHP-site | 20 fmol/mg protein |
| Density of high affinity PAA-site | 40 fmol/mg protein |
| Density of nucleoside transporter | 230 fmol/mg protein[b] |
| Density of low affinity DHP-site | 700 fmol/mg protein |

[a]Taken from Reuter (1983) and Pelzer et al. (1985).

[b]Taken from Marangos et al. (1984).

[c]The density was calculated by using a cell volume, a specific density of the cell and a protein content of 10 pl, 1.1, and 20% of wet weight. *DHP*, dihydropyridine, *PAA*, phenylalkylamine; the density of these binding sites was determined in cardiac muscle homogenates.

contrast, the number of the high affinity binding sites present in muscle homogenates (about 20-40 fmol/mg protein) is within the range of the density of the electrophysiologically determined number of channels (Table 5). This similiarity of values suggests that the sarcolemmal high affinity sites represent part of the calcium channel. Their high affinity is not in contrast to this interpretation, since Bean (1984) demonstrated recently that nitrendipine blocks the inactivated calcium channel with an apparent $K_D$ of 0.34 nM whereas the apparent $K_D$ for the resting channel is about 700 nM. It is very likely that the channel present in muscle homogenates is in its inactivated, high affinity state.

Although these considerations may be valid, their plausibility depends on the identification of the physiological function of the low affinity binding sites. Figure 2 shows that the binding of nimodipine to the sarcolemma is inhibited by 50% at nmolar concentrations of nitrobenzylthioinosine and hexobendine. The $IC_{50}$ values for these compounds are between 1-3 nM. They bind with high affinity to the low affinity site of nimodipine and do not affect binding of nimodipine to its high affinity site or the binding of (-)desmethoxyverapamil (Ruth et al. 1985). Both compounds are well characterized inhibitors of the nucleoside transporter. Nimodipine inhibits binding of radioactive nitrobenzylthioinosine in cardiac membranes with an $IC_{50}$ value of 44 nM (Marangos et al. 1984), which is identical with the $K_D$ value of 33 nM determined for the low affinity site in this study. In addition, the density of the low affinity site for nimodipine is close to that for the nucleoside transporter (Table 5), suggesting that the low affinity site for dihydropyridines is related to the nucleoside transporter and has nothing to do with the calcium channel.

## Solubilization and Partial Purification of the Sarcolemmal Binding Sites for Calcium Channel Blockers

The evidence presented so far suggests that the high affinity sites for the dihydropyridines and phenylalkylamines are located on the voltage-operated calcium channel. If this assumption is correct, one might assume that the allosteric interaction observed in the membrane between different sites is still present after solubilization of these binding sites by detergents. The sarcolemmal sites are effectively solubilized by the detergents Chaps, digitonin, and sucrose laurylmonoester. Chaps solubilizes about 40% of the membrane protein and of the binding sites, whereas the other two detergents solubilize more protein than binding sites. The high affinity site for nimodipine and (-)desmethoxyverapamil is present in the detergent extract with un-

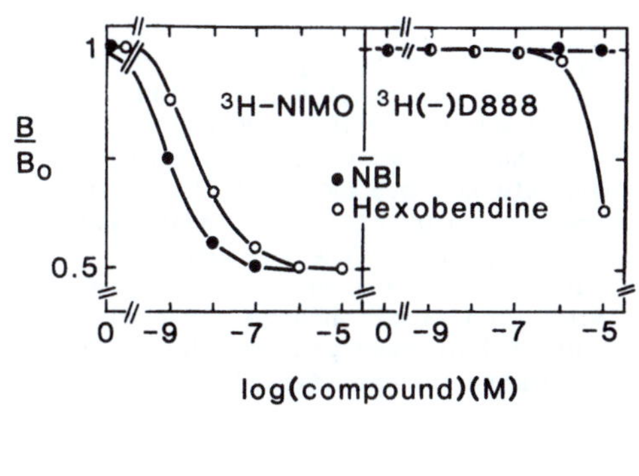


Fig. 2. Inhibitors of the nucleoside transporter inhibit binding of nimodipine. Binding of nimodipine (1 nm) and (-)desmethoxyverapamil (1 nm) was determined as desribed by Ruth et al. (1985). *NBI* nitrobenzylthioinosine

changed density and identical (nimodipine) or somewhat lower ((-)desmethoxyverapamil) affinity than before solubilization. In contrast, the affinity for the low affinity sites decreases about tenfold to $K_D$ values of 0.1 to 1 µM. The high affinity sites for nimodipine, (-)desmethoxyverapamil and d-cis-diltiazem interfere with the binding of each other in an allosteric manner after solubilization. This suggests that these sites are localized either on the same protein or on subunits of the same structure which bind tightly together.

The high affinity site for the dihydropyridines was partially purified on a Sepharose column substituted with a phenylalkylamine (Fig. 3). Binding sites are eluted with high salt and have been characterized further by using the Hummel and Dreyer technique. Equilibrium binding experiments with four different preparations yielded a $K_D$ value of 1.2 ± 0.2 nM and a density of 146 ± 4 pmol/mg. The specific density represents a 2000-fold purification of the high affinity site which may be about 10% pure. The major peptides detected by SDS gel electrophoresis had a molecular mass of 90, 58, 45, and 30 kDa. Binding of tritiated nitrendipine to this fraction was inhibited by other dihydropyridines in nmolar concentrations but only in µmolar concentrations by phenylalkylamines. The reason for the apparent loss of the allosteric interaction between the different sites is not clear, but it may be caused either by separation of these sites during purification or by a different susceptibility of these sites against proteases. The second explanation is supported by the observation that complete hydrolysis of the membrane-bound (-)desmethoxyverapamil binding site requires 1 µg trypsin/mg protein, whereas the nimodipine binding sites are destroyed completely only in the presence of 0.1 mg trypsin/mg proteins. The binding site eluted with high salt from the phenylalkylamine column was further purified on a DEAE-cellulose column. The fractions eluting between 50 and 100 mM sodium chloride bound nitrendipine. This step yielded an additional threefold purification at the expense of re-

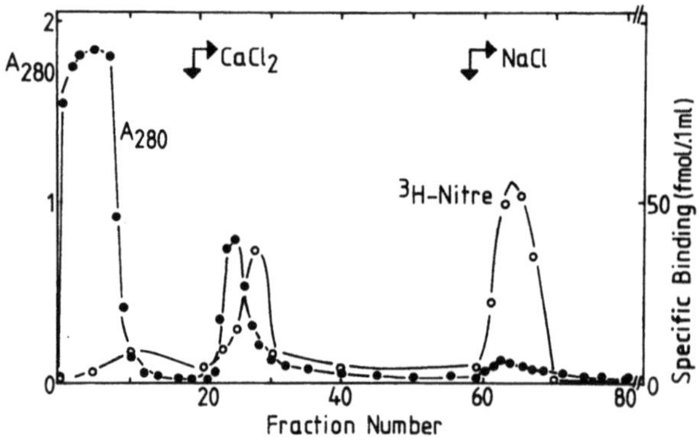


Fig. 3. Apparent purification of the high affinity site for dihydropyridines on a phenylalkylamine-Sepharose column. The hydroxy-analog of verapamil was coupled with epoxy-activated Sepharose B4 to a concentration of 12 mM. The column (30 ml) was equilibrated with buffer A (5 mM MOPS, pH 7.4, 50 mM NaCl, 5 mM Chaps, 0.5 mM EDTA, 25 µg/ml leupeptin, 10 µg/ml aprotinin, 0.5 mM benzamidine, 0.5 mM PMSF). The high speed supernatant (200 mg protein containing 0.2 pmol/mg high affinity binding sites for dihydropyridines) of crude cardiac membranes was applied to the column. The column was washed with buffer A and eluted with buffer A containing 30 mM $CaCl_2$ and 140 mM NaCl followed by buffer A containing 1 M NaCl. The fraction size was 5 ml. The binding density was determined at 4°C with 1 nM [$^3$H] nitrendipine by the Hummel and Dreyer technique using 2 µM nitrendipine to determine nonspecific binding

covered binding sites. The fraction with the highest binding density was incubated together with [$\gamma$-$^{32}$P]ATP and the catalytic subunit of cAMP-dependent protein kinase. The autoradiogram after electrophoresis in the presence of SDS showed a major phosphorylated peptide of $M_r$ 55-60 kDa. The molecular weight of this peptide is similar to that phosphorylated in the purified dihydropyridine binding site from the T-tubule of skeletal muscle (Curtis and Catterall 1984).

## Conclusions

The studies carried out so far provide evidence that:

a) catecholamines increase the calcium conductance of cardiac muscle cells by cAMP-dependent phosphorylation of the voltage-operated calcium channel or a closely related protein;
b) cardiac muscle membranes contain several binding sites for calcium channel blockers which are unrelated to the calcium channel;
c) the sarcolemmal high affinity sites for the dihydropyridines and the phenylalkylamines are part of the voltage-operated calcium channel;
d) the calcium channel contains three distinct sites for the three groups of calcium channel blockers which interact with each other by an allosteric mechanism;
e) a peptide of 55 to 60 kDa may be involved in the regulation of the calcium channel by cAMP-dependent phosphorylation.

These properties of the cardiac calcium channel are partially reminiscent of the properties of the calcium channel blocker binding sites present in skeletal T-tubule membranes. However, so far an identity of these two calcium channels has not been established since a completely purified preparation of a mammalian cardiac calcium channel is not available. The peptides present in the partially purified cardiac preparation are similar to that reported to be present in a purified calcium channel from chicken heart (Rengasamy et al. 1985). This raises the possibility that the cardiac calcium channel differs from that present in skeletal T-tubule. This conclusion is supported by the fact that the physiological significance of these two calcium channels is quite different.

*Acknowledgments.* We would like to thank Mrs. Vogt for expert technical assistance, Mrs. Siepmann for typing the manuscript, Dr. Hoffmeister from Bayer AG, Dr. Hollmann and Dr. Kretzschmar from Knoll AG, and Dr. Satzinger and Dr. Marmê from Gödecke AG for providing some of the organic calcium channel blockers used in this study. This work was supported by grants grom DFG and Fonds der Chemischen Industie.

## References

Armstrong CM, Matteson DR (1985) Science 227:65-67
Bartschat DK, Cyr DL, Lindenmayer GE (1980) J Biol Chem 255:10044-10047
Bean BP (1984) Proc Natl Acad Sci USA 81:6388-6392
Carbone E, Lux HD (1984) Nature (London) 310:501-502
Curtis BM, Catterall WA (1984) Biochemistry 23:2113-2118
Fleckenstein A (1977) Annu Rev Pharmacol Toxicol 17:149-166
Flockerzi V, Mewes R, Ruth P, Hofmann F (1983) Eur J Biochem 135:131-142
Glossmann H, Ferry D, Lübbecke F, Mewes R, Hofmann F (1982) Trends Pharmacol Sci 3:431-437

Glossmann H, Ferry D, Goll A (1985) Proc IX Int Congr Pharmacol London 2:329-335
Hess P, Lausman JB, Tsien RW (1984) Nature (London) 311:538-544
Hofmann F, Bechtel PJ, Krebs EG (1977) J Biol Chem 252:1441-1447
Jones LR, Cala SE (1981) J Biol Chem 256:11009-11018
Kameyama M, Hofmann F, Trautwein W (1985) Pflüger's Arch 405:285-293
Marangos PJ, Finkel MS, Verma A, Maturi MF, Patel J, Patterson RE (1984) Life Sci 35:1109-1116
Millard RW, Grupp G, Grupp IZ, Di Salvo J, De Pover A, Schwartz A (1983) Circ Res 52:Suppl I, 29-39
Murphy KMM, Gould RJ, Largent BL, Snyder S (1983) Proc Natl Acad Sci USA 80:860-864
Osterrieder W, Brum G, Hescheler J, Trautwein W, Flockerzi V, Hofmann F (1982) Nature (London) 298:576-578
Pelzer D, Hescheler J, Cavalie A, Trautwein W (1985) Pfluegers Arch 403:Suppl R 48
Rengasamy A, Ptasinski J, Hosey MM (1985) Biochem Biophys Res Commun 126:1-7
Reuter H (1983) Nature (London) 569-574
Reuter H (1984) Annu Rev Physiol 46:473-484
Rinaldi ML, Le Peuch CJ, Demaille JG (1982) FEBS Lett 129:277-281
Ruth P, Flockerzi V, v. Nettelbladt E, Oeken J, Hofmann F (1985) Eur J Biochem 150: 313-322

# Phosphorylation of Ion Channels: A Fundamental Regulatory Mechanism in the Control of Nerve Cell Activity

I. B. Levitan[1]

## Introduction

There has been much progress in recent years in our understanding of the molecular mechanisms by which neurotransmitters exert their effects on excitable cells. On the one hand, studies of the structure and function of the nicotinic acetylcholine receptor have provided a detailed picture of a *directly coupled* receptor/channel system, a system in which the opening of an ion channel (and transport of ions) is dependent on the continued occupation of a closely associated receptor by the transmitter. However, there are many examples of physiological responses which outlast the initial stimulus, the occupancy of the receptor by the transmitter, by seconds, minutes, or even hours. It is difficult to explain such long-lasting effects in terms of direct receptor/channel coupling, and it seems more likely that it results from some long-lasting metabolic modification of the channel. In this case the receptor and channel may not necessarily be intimately associated in a single macromolecular complex, but may be *indirectly coupled* via some intracellular second messenger which is produced upon occupancy of receptor by neurotransmitter. The second messenger sets in motion a series of steps, culminating in some covalent modification of the channel which alters its activity, and the functional change persists until the covalent modification has been reversed (again this may require a series of steps).

The search for second messengers which might mediate long-lasting modulation of ion channel activity in excitable cells has drawn heavily on studies from non-neuronal systems. Cyclic AMP (cAMP) and calcium ions are well established as second messengers in the regulation of carbohydrate metabolism in such tissues as liver and muscle, where they modulate the activity of certain enzymes by causing them to be phosphorylated (for a review see Cohen 1982). It has long been suspected that they might play a similar role in the regulation of ion channels, and recent evidence from several laboratories has confirmed that this is indeed the case. This paper will focus on some of the experimental strategies which have been and are being used to investigate the modulation of ion channels in excitable cells by calcium, cAMP, and cAMP-dependent protein phosphorylation.

## A Second Messenger Role for cAMP in Nerve Cells

If one wants to determine whether a particular physiological response is mediated by cAMP, one obvious question to ask is whether cAMP can mimic the response. With this in mind, a large number of investigators have applied cAMP to cells while monitoring their membrane properties

[1] Graduate Department of Biochemistry, Brandeis University, Waltham, MA 02254, USA

36. Colloquium - Mosbach 1985
Neurobiochemistry

(for a review see Siegelbaum and Tsien 1983). In some cases membrane-permeable derivatives of cAMP have been applied in the extracellular medium, while in others cAMP itself or derivatives which are resistant to hydrolysis by phosphodiesterases have been injected intracellularly via microelectrodes. In other experiments, intracellular cAMP levels have been elevated by application of phosphodiesterase inhibitors, which inhibit the breakdown of the cyclic nucleotide, or activators of adenylate cyclase, which increase its synthesis. Results from all of these approaches indicate that cAMP can indeed modulate membrane properties. In some of these studies, voltage clamp analysis has allowed the identification of particular ionic currents which can be affected by cAMP. There appears to be no single ion current which is a universal target for cAMP, but rather the current which is modulated is different from cell to cell (Siegelbaum and Tsien 1983).

A demonstration of a *pharmacological* effect of cAMP on membrane properties is, of course, not in itself evidence that cAMP plays a *physiological* role. However, in several of the studies referred to above, it has been shown that cAMP mimics the action of a particular physiological agonist, and that the physiological agonist can cause the stimulation of adenylate cyclase and the accumulation of cAMP in the target cell. These and other pharmacological, biochemical, and electrophysiological experiments have indeed provided strong circumstantial evidence for a physiological role for cAMP in some cells. Among the physiological responses which appear to be mediated by cAMP are the activation of the slow inward calcium current by β-adrenergic stimulation of cardiac cells, and the inactivation or activation of several distinct potassium currents by serotonin in several different molluscan neurons. One example taken from work in our laboratory is illustrated in Fig. 1. Serotonin causes the identified *Aplysia* neuron R15 to hyperpolarize (Drummond et al. 1980), and as shown in Fig. 1, this is accompanied by an increase in the slope of the steady-state current-voltage curve at hyperpolarized potentials. A series of ion substitution and pharmacological blocking experiments have established that this is due to the activation of an inwardly rectifying potassium channel (Benson and Levitan 1983). Furthermore, cAMP mimics this response (Fig. 1), and serotonin activates adenylate cyclase and causes cAMP to accumulate within R15 (Levitan 1978). These and other combined biochemical and physiological experiments have provided strong evidence that the serotonin-evoked increase in the activity of this potassium channel is mediated by cAMP. Similar data are now available for several other molluscan neurons.

Convincing evidence for second messenger involvement in transmitter-evoked physiological responses in neurons has been surprisingly difficult to come by. As pointed out above, there are now several cases for which the evidence is becoming quite compelling, but much of it is indirect. One difficulty is that no general criteria have been available to identify second messenger coupling, independent of what the messenger might be. However, appropriate use of recently developed patch recording techniques does provide such a general test.

Since a gigohm seal between a patch recording electrode and a membrane forms a lateral diffusion barrier, it follows that neurotransmitter applied outside the electrode cannot activate receptors in the patch of membrane inside the electrode. Thus, if channels inside a patch are affected by transmitter applied outside, this must be due to occupancy of receptors located outside, and these receptors can communicate with the channels inside the patch only by means of a diffusible intracellular messenger. In contrast, if a receptor is coupled directly to an ion channel (e.g., the nicotinic acetylcholine receptor/channel), it

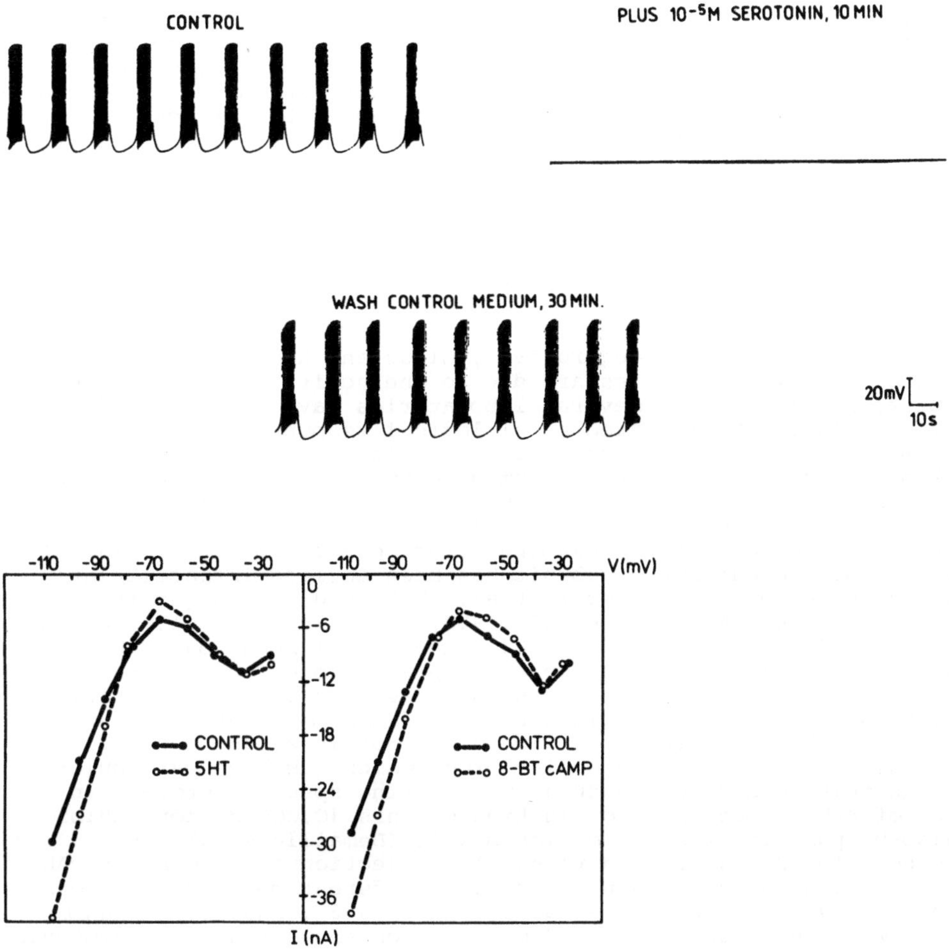


Fig. 1. Effects of serotonin (5HT) and 8-benzylthio cAMP (8-BT cAMP) on the electrical properties of *Aplysia* neuron R15. The *top set of traces* are recordings of the normal bursting pattern of electrical activity in neuron R15, the hyperpolarization and inhibition of bursting produced by adding serotonin to the bathing medium, and the return to the normal bursting pattern after washing out the serotonin. The *bottom* shows steady-state current-voltage curves measured under voltage clamp. Serotonin elicits an increase in the slope of the curve, indicative of an increase in K+ conductance. This effect can be mimicked by 8-BT cAMP. (After Drummond et al. 1980)

should be possible to affect a channel in the electrode only by applying transmitter inside.

This general test has been used in at least two laboratories to identify second messenger-mediated responses. Siegelbaum et al. (1982), working with sensory neurons from the abdominal ganglion of *Aplysia*, have demonstrated that serotonin applied in the extracellular medium can cause the closure of individual potassium channels inside a patch electrode. Other experiments have shown that the second messenger which mediates this response is cAMP (but this knowledge was not necessary for the demonstration that *some* second messenger must be involved) Maruyama and Petersen (1982) have used a similar approach to show that cholecystokinin and acetylcholine activation of single calcium-dependent cation channels in pancreatic acinar cells (which may be considered honorary neurons for our purposes) is mediated by intracellular calcium ions. These elegant experiments have introduced a new level of sophistication to the way we think about and study second messenger-mediated physiological responses in excitable cells.

## Protein Phosphorylation Affects Membrane Properties

It is thought (Kuo and Greengard 1969, Glass and Krebs 1980) that all actions of cAMP in eukaryotes are due to the activation of cAMP-dependent protein kinases, and several laboratories have investigated the possibility that cAMP-mediated physiological responses involve protein phosphorylation. The active catalytic subunit of cAMP-dependent protein kinase has been purified to homogeneity from several mammalian sources (Peters et al. 1977). Its active site has been well conserved during evolution, and thus it is possible to use the mammalian enzyme as a pharmacological tool to examine membrane excitability in mammals and molluscs. The catalytic subunit has been applied intracellularly in cardiac myocytes (Osterrieder et al. 1982) and several different molluscan neurons (Kaczmarek et al. 1980, Castellucci et al. 1980, De Peyer et al. 1982, Alkon et al. 1983), and it has been found that it can affect membrane properties in a way predictable from the known actions of cAMP and neurotransmitters on these same cells. A complementary approach to the pharmacological use of *exogenous* catalytic subunit is to ask whether inhibition of *endogenous* protein kinase can affect a physiological response. A molecular probe which has been particularly useful in this regard is a naturally occurring specific protein inhibitor of cAMP-dependent protein kinase. This 10,000 dalton inhibitor can also be purified from mammalian muscle (Demaille et al. 1977), and it has been found that its intracellular injection can block the cAMP-mediated actions of serotonin in several different molluscan neurons (Adams and Levitan 1982, Castellucci et al. 1982). Again, an example from our work is illustrated in Fig. 2, which shows that the serotonin-evoked increase in the inwardly rectifying potassium channel in neuron R15 is blocked by the injection of the protein kinase inhibitor.

## Physiologically Relevant Phosphorylated Proteins

Once it is demonstrated that cAMP-dependent protein phosphorylation can modulate neuronal electrical activity, it of course becomes very important to identify those phosphorylated proteins which are involved in this modulation. We recently developed a technique to examine phosphorylation in a single living neuron following the intracellular injection of radioactive ATP (Lemos et al. 1982, 1984), under conditions which allow the simultaneous monitoring of the cell's electrical properties using voltage clamp. A particularly important feature of this in vivo single cell technique is that the cell's "physiological history" during the labeling period can be compared directly with the phosphoprotein labeling pattern, to determine whether alterations in the phosphorylation of specific proteins are related to a particular physiological response. Using this technique, we have identified a number of phosphoproteins in neuron R15 whose phosphorylation state

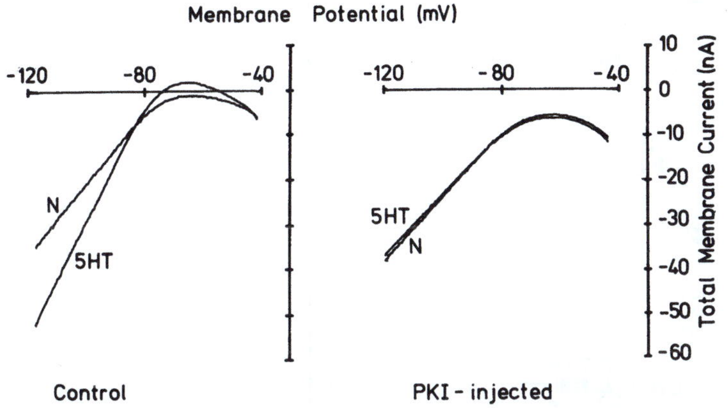


Fig. 2. Inhibition of the serotonin effect on neuron R15 by protein kinase inhibitor (*PKI*). The normal serotonin (*5HT*) elicited increase in K+ conductance in neuron R15 (*left*) is absent following the intracellular injection of the 10,000 dalton PKI (*right*), indicating that cAMP-dependent protein phosphorylation is a *necessary* step in this action of serotonin. (After Adams and Levitan 1982)

is altered by serotonin, concomitant whith the serotonin-evoked increase in the activity of the inwardly rectifying potassium channel in this cell (Lemos et al. 1982, 1984). More recently, we have examined the pharmacology and kinetics of these phosphorylation changes to determine whether all or only some are tightly correlated with the potassium conductance change. Using this approach, we have identified two phosphoproteins, of molecular weight 29,000 and 70,000 daltons (Fig. 3), whose phosphorylation cannot be dissociated from the increase in potassium conductance whatever the experimental manipulation (Lemos et al. 1985). Thus, these phosphoproteins are candidates to be ion channel proteins or channel regulatory components, and future work will be directed towards purifying these phosphoproteins and characterizing their role in channel modulation.

## Phosphorylation of Ion Channels

The approaches described above have provided strong evidence that cAMP-dependent protein phosphorylation is a necessary step in the actions of some neurotransmitters on membrane properties, and have made considerable progress in identifying some phosphoproteins which may play a role in long-term neuromodulation. However, they do not make clear whether it is the ion channel itself which is phosphorylated, or whether phosphorylation is simply an early step in some cascade of events which leads ultimately to modulation of channel activity. Again, single channel recording techniques have provided a way to examine this question. The gigohm seal between an electrode and a membrane patch is not only electrically tight, it is also mechanically tight, and thus, the patch can be pulled free from the cell under conditions which result in the cytoplasmic face of the membrane being exposed to the bathing medium (Hamill et al. 1981). When such isolated patches from *Aplysia* (Shuster et al. 1985) or *Helix* (Ewald et al. 1985) neurons are bathed in catalytic subunit and ATP, there is a change in the activity of individual ion channels in the patch. Again, an example from our work is shown in Fig. 4, which demonstrates that the activity of

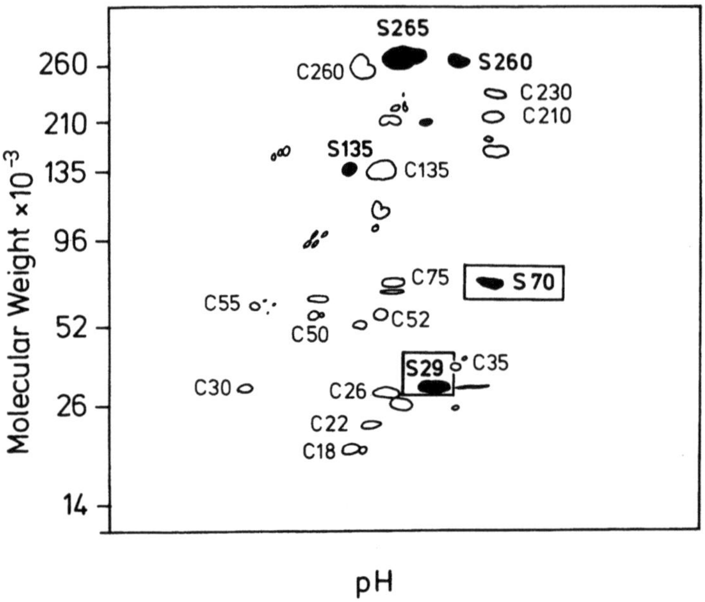


Fig. 3. Phosphorylated proteins associated with the serotonin-elicited increase in K+ conductance in neuron R15. Schematic representation of an autoradiogram of a two-dimensional polyacrylamide gel on which radioactive phosphoproteins from neuron R15 have been separated. Following the injection of radioactive ATP into R15, a number of radioactive phosphoproteins can be observed (some are marked with a *C* for control and their molecular weight on the figure). In a serotonin-treated cell, a number of additional phosphoproteins (marked with an *S* and their molecular weight) are labeled. Most of these can be dissociated from the K+ conductance increase by one or another pharmacological manipulation; however, the phosphoproteins labeled *S29* and *S70* (*boxes*) appear to be closely associated with the physiological response. (After Lemos et al. 1985)

calcium-dependent potassium channels in patches from *Helix* neurons can be modulated by cAMP-dependent protein phosphorylation. These results indicate that the target for phosphorylation must be some protein which remains associated with the patch of membrane when it is pulled away from the cell. This target might be the channel itself, or it might be some cytoskeletal component or other protein which comes away with the patch.

A related approach involves the examination of ion channel activity in a reconstituted system. It is possible to fuse membrane vesicles, containing functional ion channels, with artificial lipid bilayers under conditions which allow the current passing through individual ion channels in the bilayer to be measured (Miller 1983). In fact, the bilayer can even be formed on the tip of a patch recording electrode (Wilmsen et al. 1983); this marriage between the patch and reconstitution methods provides single channel recordings with particularly favorable signal-to-noise characteristics because of the small size of the lipid bilayer. When calcium-dependent potassium channels from *Helix* neurons are examined in these reconstituted systems, addition of catalytic subunit and ATP to one side of the bilayer results in an increase in the activity of the individual channels (Fig. 5). Since proteins are essentially at infinite dilution in such planar bilayers, it can be con-

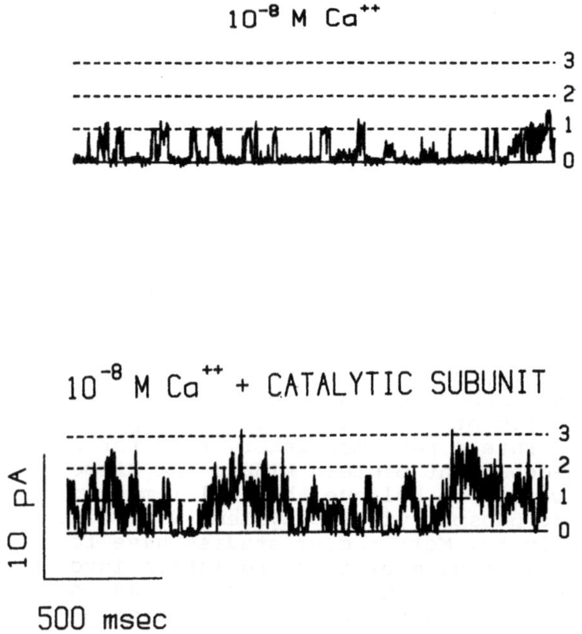


Fig. 4. Effect of the catalytic subunit of cAMP-dependent protein kinase on the activity of $Ca^{2+}$-dependent $K^+$ channels in isolated membrane patches from *Helix* neurons. The *top trace* shows the activity of channels in a detached patch at low calcium concentration. From records taken at higher calcium concentration (not shown) it was determined that there were at least 4 $K^+$ channels in this patch and that their gating was calcium-dependent. The bottom trace shows that the channels were much more active, even at the low calcium concentration, after addition of the catalytic subunit together with Mg++ and ATP. These results indicate that phosphorylation of some protein which comes away with the detached patch gives rise to altered activity of ion channels in the patch. (After Ewald et al. 1985)

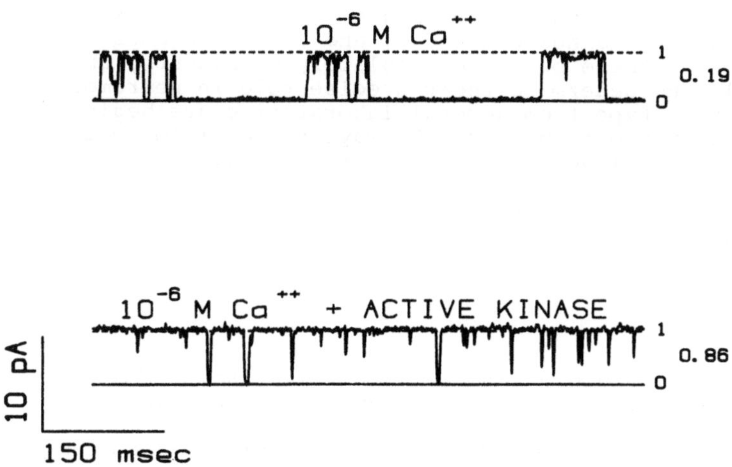


Fig. 5. Effect of the catalytic subunit of cAMP-dependent protein kinase on the activity of a $Ca^{2+}$-dependent $K^+$ channel reconstituted in an artificial phospholipid bilayer. Membrane vesicles from *Helix* neurons were reconstituted by freezing and thawing with exogenous phospholipid, and a bilayer was formed on the tip of a patch recording electrode (Wilmsen et al. 1983). The *top trace* shows the activity of a single $K^+$ channel in this bilayer at a voltage of +40 mV. Following the addition of the catalytic subunit together with $Mg^{2+}$ and ATP, the activity of the channel changed dramatically (*bottom trace*). The magnitude of the change in activity induced by phosphorylation is illustrated by the *numbers to the right* of each trace, which indicate the proportion of time the channel was open under each experimental condition. These results indicate that the phosphorylation target must be closely associated with the ion channel protein. (After Ewald et al. 1985)

cluded that the phosphorylation target must be either the channel itself or something intimately associated with the channel.

## Summary and Conclusions

It is no accident that many of the major advances in the field of neuromodulation have come from studies on molluscan neurons; their large size and ready identifiability allow combined biochemical, pharmacological, and electrophysiological experiments to be carried out on individual neurons, and such an interdisciplinary approach is certainly necessary for a complete understanding of long-term neuromodulation. Work along these lines from several laboratories has demonstrated convincingly that the electrical activity of some neuronal and cardiac cells can be regulated by cAMP and cAMP-dependent protein phosphorylation (for a recent review see Levitan et al. 1983). Similar approaches have recently been used to demonstrate a neuromodulatory role for C kinase, the calcium/phospholipid-dependent protein kinase (De Riemer et al. 1985), and there is good reason to believe that calcium/calmodulin-dependent and cGMP-dependent phosphorylating systems will also be shown to regulate neuronal activity. More recent studies have focused on the possibility that the mechanism of this regulation involves the direct phosphorylation of ion channels in the plasma membrane of these excitable cells. This possibility is supported by the single channel recording experiments, which demonstrate the modulation by phosphorylation of individual ion channels, both in isolated membrane patches and reconstituted in artificial phospholipid bilayers. These results in turn provide an indication that one or more of the radioactive phosphoproteins identified in the single cell labeling experiments may indeed be an ion channel component, and it now becomes essential to attempt to purify these putative regulatory components, characterize them in detail, and determine their precise role in neuromodulation. Experiments of this type from several laboratories are beginning to provide fundamental information about the ways neuronal ion channels, and hence neuronal electrical activity, can be regulated.

*Acknowledgments.* This work was supported by Grant NS17910 from the National Institute of Neurological and Communicative Disorders and Stroke.

## References

Adams WB, Levitan IB (1982) Intracellular injection of protein kinase inhibitor blocks the serotonin-induced increase in $K^+$ conductance in *Aplysia* neuron R15. Proc Natl Acad Sci USA 79:3877-3880

Alkon DL, Acosta-Urquidi J, Olds J, Kuzma G, Neary JT (1983) Protein kinase injection reduces voltage-dependent potassium currents. Science 219:303-306

Benson JA, Levitan IB (1983) Serotonin increases an anomalously rectifying $K^+$ current in the *Aplysia* neuron R15. Proc Natl Acad Sci USA 80:3522-3525

Castellucci VF, Kandel ER, Schwartz JH, Wilson FD, Nairn AC, Greengard P (1980) Intracellular injection of the catalytic subunit of cyclic AMP-dependent protein kinase simulates facilitation of transmitter release underlying behavioral sensitization in *Aplysia*. Proc Natl Acad Sci USA 77:7492-7496

Castellucci VF, Nairn A, Greengard P, Schwartz JH, Kandel ER (1982) Inhibitor of Adenosine 3':5'-monophosphate-dependent protein kinase blocks presynaptic facilitation in *Aplysia*. J Neurosci 2:1673-1681

Cohen P (1982) The role of protein phosphorylation in neural and hormonal control of cellular activity. Nature (London) 296:613-620

Demaille J, Peters K, Fischer E (1977) Isolation and properties of the rabbit skeletal muscle protein inhibitor of cAMP-dependent protein kinases. Biochemistry 16:3080-3086

De Peyer JE, Cachelin AB, Levitan IB, Reuter H (1982) $Ca^{2+}$-activated $K^+$ conductance in internally perfused snail neurons is enhanced by protein phosphorylation. Proc Natl Acad Sci USA 79:4207-4211

De Riemer SA, Strong JA, Albert KA, Greengard P, Kaczmarek LK (1985) Enhancement of calcium current in *Aplysia* neurones by phorbol ester and protein kinase C. Nature (London) 313:313-316

Drummond AH, Benson JA, Levitan IB (1980) Serotonin-induced hyperpolarization of an identified *Aplysia* neuron is mediated by cyclic AMP. Proc Natl Acad Sci USA 77: 5013-5017

Ewald D, Williams A, Levitan IB (1985) Modulation of single $Ca^{++}$-dependent $K^+$ channel activity by protein phosphorylation. Nature 315:503-506

Glass DB, Krebs EG (1980) Protein phosphorylation catalyzed by cyclic AMP-dependent and cyclic GMP-dependent protein kinases. Annu Rev Pharmacol Toxicol 20:363-388

Hamill OP, Marty A, Neher E, Sakmann B, Sigworth FJ (1981) Improved patch-clamp techniques for high-resolution current recording from cells and cell-free membrane patches. Pfluegers Arch 391:81-100

Kaczmarek LK, Jennings KR, Strumwasser F, Nairn AC, Walter U, Wilson FD, Greengard P (1980) Microinjection of catalytic subunit of cyclic AMP-dependent protein kinase enhances calcium action potentials of bag cell neurons in cell culture. Proc Natl Acad Sci USA 77:7487-7491

Kuo JF, Greengard P (1969) Cyclic nucleotide-dependent protein kinases IV. Widespread occurrence of adenosine 3',5'-monophosphate-dependent protein kinase in various tissues and phyla of the animal kingdom. Proc Natl Acad Sci USA 64:1359-1365

Lemos JR, Novak-Hofer I, Levitan IB (1982) Serotonin alters the phosphorylation of specific proteins inside a single living nerve cell. Nature (London) 298:64-65

Lemos JR, Novak-Hofer I, Levitan IB (1984) Synaptic stimulation alters protein phosphorylation in vivo in a single *Aplysia* neuron. Proc Natl Acad Sci USA 81:3233-3237

Lemos JR, Novak-Hofer I, Levitan IB (1985) Phosphoproteins associated with the regulation of a specific potassium channel in the identified *Aplysia* neuron R15. J Biol Chem 260:3207-3214

Levitan IB (1978) Adenylate cyclase in isolated *Helix* and *Aplysia* neuronal cell bodies: stimulation by serotonin and peptide-containing extract. Brain Res 154: 404-408

Levitan IB, Lemos JR, Novak-Hofer I (1983) Protein phosphorylation and the regulation of ion channels. Trends Neurosciences 6:496-499

Maruyama Y, Petersen OH (1982) Cholecystokinin activation of single-channel currents is mediated by internal messenger in pancreatic acinar cells. Nature (London) 300:61-63

Miller C (1983) Integral membrane channels: studies in model membranes. Physiol Rev 63:1209-1242

Osterrieder W, Brum G, Hescheler J, Trautwein W, Flockerzi V, Hofmann F (1982) Injection of subunits of cyclic AMP-dependent protein kinase into cardiac myocytes modulates $Ca^{2+}$ current. Nature (London) 298:576-578

Peters KA, Demaille JG, Fischer EH (1977) Adenosine 3':5'-monophosphate-dependent protein kinase from bovine heart. Characterization of the catalytic subunit. Biochemistry 16:5691-5697

Shuster M, Camardo J, Siegelbaum S, Kandel ER (1985) Cyclic AMP-dependent protein kinase closes the serotonin-sensitive $K^+$ channels of *Aplysia* sensory neurones in cell-free membrane patches. Nature (London) 313:392-395

Siegelbaum SA, Tsien RW (1983) Modulation of gated ion channels as a mode of transmitter action. Trends Neurosci 6:307-313

Siegelbaum SA, Camardo JS, Kandel ER (1982) Serotonin and cyclic AMP close single $K^+$ channels in *Aplysia* sensory neurones. Nature (London) 299:413-417

Wilmsen U, Methfessel C, Hanke W, Boheim G (1983) Channel current fluctuation studies with solvent-free lipid bilayers using Neher-Sakmann pipettes. In: Physical chemistry of transmembrane ion motions. Elsevier, Amsterdam, pp 479-485

# Subject Index

MIX
Papier aus verantwortungsvollen Quellen
Paper from responsible sources
FSC® C105338

If you have any concerns about our products,
you can contact us on
**ProductSafety@springernature.com**

In case Publisher is established outside the EU,
the EU authorized representative is:
**Springer Nature Customer Service Center GmbH**
**Europaplatz 3, 69115 Heidelberg, Germany**

Printed by Libri Plureos GmbH
in Hamburg, Germany